Pigs, Pork, and Heartland Hogs

Rowman & Littlefield Studies in Food and Gastronomy

Food studies is a vibrant and thriving field encompassing not only cooking and eating habits but also issues such as health, sustainability, food safety, and animal rights. Scholars in disciplines as diverse as history, anthropology, sociology, literature, and the arts focus on food. The mission of **Rowman & Littlefield Studies in Food and Gastronomy** is to publish the best in food scholarship, harnessing the energy, ideas, and creativity of a wide array of food writers today. This broad line of food-related titles will range from food history, interdisciplinary food studies monographs, general interest series, and popular trade titles to textbooks for students and budding chefs, scholarly cookbooks, and reference works.

Appetites and Aspirations in Vietnam: Food and Drink in the Long Nineteenth Century, by Erica J. Peters
Three World Cuisines: Italian, Mexican, Chinese, by Ken Albala
Food and Social Media: You Are What You Tweet, by Signe Rousseau
Food and the Novel in Nineteenth-Century America, by Mark McWilliams
Man Bites Dog: Hot Dog Culture in America, by Bruce Kraig and Patty Carroll
A Year in Food and Beer: Recipes and Beer Pairings for Every Season, by Emily Baime and Darin Michaels
Celebraciones Mexicanas: History, Traditions, and Recipes, by Andrea Lawson Gray and Adriana Almazán Lahl
The Food Section: Newspaper Women and the Culinary Community, by Kimberly Wilmot Voss
Small Batch: Pickles, Cheese, Chocolate, Spirits, and the Return of Artisanal Foods, by Suzanne Cope
Food History Almanac: Over 1,300 Years of World Culinary History, Culture, and Social Influence, by Janet Clarkson
Cooking and Eating in Renaissance Italy: From Kitchen to Table, by Katherine A. McIver
Eating Together: Food, Space, and Identity in Malaysia and Singapore, by Jean Duruz and Gaik Cheng Khoo
Nazi Hunger Politics: A History of Food in the Third Reich, by Gesine Gerhard
The Carrot Purple and Other Curious Stories of the Food We Eat, by Joel S. Denker
Food in the Gilded Age: What Ordinary Americans Ate, by Robert Dirks
Urban Foodways and Communication: Ethnographic Studies in Intangible Cultural Food Heritages Around the World, by Casey Man Kong Lum and Marc de Ferrière le Vayer
Food, Health, and Culture in Latino Los Angeles, by Sarah Portnoy
Food Cults: How Fads, Dogma, and Doctrine Influence Diet, by Kima Cargill
Prison Food in America, by Erika Camplin
K'Oben: 3,000 Years of the Maya Hearth, by Amber M. O'Connor and Eugene N. Anderson
As Long As We Both Shall Eat: A History of Wedding Food and Feasts, by Claire Stewart
American Home Cooking: A Popular History, by Tim Miller
A Taste of Broadway: Food in Musical Theater, by Jennifer Packard
Pigs, Pork, and Heartland Hogs: From Wild Boar to Baconfest, by Cynthia Clampitt
Sauces Reconsidered: Après Escoffier, by Gary Allen
Nomadic Food, by Jean Pierre Williot and Isaelle Bianquis
Screen Cuisine: Food and Film from Prohibition to James Bond, by Linda Civitello
The Social Implications of Vegetarian Food Choices: A Narrative Approach to Experiences Around the Globe, edited by Charlotte De Backer; Julie Dare and Leesa Costello

Pigs, Pork, and Heartland Hogs

From Wild Boar to Baconfest

Cynthia Clampitt

ROWMAN & LITTLEFIELD
Lanham • Boulder • New York • London

Published by Rowman & Littlefield
An imprint of The Rowman & Littlefield Publishing Group, Inc.
4501 Forbes Boulevard, Suite 200, Lanham, Maryland 20706
www.rowman.com

Unit A, Whitacre Mews, 26-34 Stannary Street, London SE11 4AB

British Library Cataloguing in Publication Information Available

Library of Congress Cataloging-in-Publication Data

Names: Clampitt, Cynthia, author.
Title: Pigs, pork, and Heartland hogs : from wild boar to baconfest / Cynthia Clampitt.
Description: Lanham : Rowman & Littlefield, [2018] | Series: Rowman & Littlefield studies in food and gastronomy | Includes bibliographical references and index.
Identifiers: LCCN 2018008925 (print) | LCCN 2018020323 (ebook) | ISBN 9781538110751 (Electronic) | ISBN 9781538110744 (cloth : alk. paper)
Subjects: LCSH: Swine—United States—History. | Pork—United States—History. Classification: LCC SF395.8.A1 (ebook) | LCC SF395.8.A1 C53 2018 (print) | DDC 636.4—dc23
LC record available at https://lccn.loc.gov/2018008925

∞™ The paper used in this publication meets the minimum requirements of American National Standard for Information Sciences—Permanence of Paper for Printed Library Materials, ANSI/NISO Z39.48-1992.

Printed in the United States of America

Contents

Section III: Living with Pigs Today

Acknowledgments

This book would not have turned out nearly so well without the generous assistance of friends from the world of food, farming, and academe. These folks sent articles, introduced me to experts, directed me toward resources, tested my recipes, or in some other way aided the process. For these helpful kindnesses and others, thanks to Catherine Lambrecht, Gil J. Stein, Melissa Ramer, Jeudi Jeuten, Cindy Ladd, Dan Sabers, Barbara Hague, Rae Eighmey, Karyn Saemann, Sylvia Zimmerman, Jill Silva, Terese Allen, Ellen Steinberg, Grace Pedalino, Carol Lopriore, Ella Strahammer, Lisa Hawkins, Louise Burton, and Cindy Cunningham.

Warmest thanks to all the farmers, researchers, historians, educators, and other experts who were so generous with their time, information, insights, and stories.

Thanks to Ken Albala, for his willingness to consider the book for his food series, and to Suzanne Staszak-Silva at Rowman & Littlefield for her enthusiasm for the project.

Thanks to Teresa Tucker for feedback on the early manuscript and to Ro Sila for the many ways she has contributed to this and other books.

For supplying photographs as well as information, my thanks to Ron Birkenholtz of the Iowa Pork Producers Association and Rick Williamson at Hormel, and for the pork-cuts graphic, my thanks to Craig "Meathead" Goldwyn at AmazingRibs.com.

To the staff of the Interlibrary Loan Department at Arlington Heights Memorial Library: I can't tell you how much I appreciate your help.

And, finally, thanks to Lizzy, Becca, Scott, Mallery, Kevin, Krissy, Jillian, Steve, Robert, Israel, and Amanda at AH Starbucks. Always felt welcome at my "second office."

Introduction

"If you want a subject, look to Pork!"

—*Great Expectations*, by Charles Dickens

In Dickens's tale, at the prompting of Mr. Pumblechook, conversation suddenly turns from happy commentary on the delights of the plump, juicy pork that diners had been enjoying to an impromptu lecture from Mr. Wopsle on the evils of swinish behavior. "The gluttony of Swine is put before us, as an example to the young," Mr. Wopsle reflects, adding that a quality "detestable in a pig is more detestable in a boy." And so Dickens encapsulates the dichotomy of attitudes toward pigs—they are both desirable and problematic. Thus it has been for most of the history of these creatures, though they have been both more desirable and more problematic for different people at different times.

Why Talk about Pigs?

Regardless of how individuals or societies feel about pigs, the fact remains that pigs provide the most commonly eaten meat in the world.[1] In fact, there are more than one billion domesticated pigs being raised worldwide,[2] plus stunning numbers of wild swine. This is true in spite of the fact that major population groups ban consumption of pork. Reliance on pigs and their progenitors stretches far back into prehistory, starting soon after the great glaciers began to recede, and, today, pretty much spans the globe.

Pigs were ideal food animals, especially as civilizations began to emerge. They are prolific (only rabbits breed faster), grow quickly, are tasty, and eat anything. On top of this, the remarkable bonus that pigs offered was that the meat can be preserved relatively easily with smoke and/or salt, both of which came into use in the misty ages of prehistory. Plus, pigs provided a ready source of fat (lard) that expanded the range of food preparation options but also offered the tremendous advantage of not needing refrigeration—still an important feature in many parts of the world.

In Europe, for more than two thousand years, pork was virtually the only meat available to peasants—though the middle and upper classes raised pigs, too. In China, if one asks simply for "meat," pork will appear. Because of their tremendous cultural and culinary importance, pigs were carried everywhere during the Age of Exploration.

Pigs readily adapted to life in the Americas. They were cherished by Conquistadors and were generally a boon in what would become Latin America, where protein had always been in short supply. Britain's North American colonies and later the United States also relied on pigs, along with the traditions and recipes that traveled with them. (Though Native Americans did not always have a good opinion of these new animals.)

In North America, particularly the part that would become the United States, the story of pork is inextricably linked with the story of corn. Corn saved early settlers, but before long, "hogs and hominy" became an aphorism for an American agricultural paradigm and the abundance it produced. It even became a temporary state motto for Tennessee. Today, the state that raises the most corn (Iowa) also raises the most pigs.

So, pigs are a very big story.

Why So Much Focus on the Midwest?

Though often overlooked in discussions of the regions of the United States, the Middle West is absolutely vital to the country's survival. In an 1862 speech, Abraham Lincoln said of the Heartland, "This great interior region is naturally one of the most important in the world."[3] The North won the Civil War at least in part because the Midwest was able to feed the Union Army, in addition to supplying a high percentage of the soldiers who fought the war.[4] Food from the Midwest has enabled the winning of overseas conflicts dating from the Crimean War in the 1800s through two World Wars and up to the present. It has also made it possible for the United States to help her allies around the world during food shortages and disasters. The Midwest is not called the Heartland simply because it is at the nation's center. It truly is the heart of the nation, both producing and pumping much of the nation's economic blood.

While corn is not all that the region produces, a considerable part of the Midwest coincides with the region known traditionally as the Corn Belt. As noted above, where hominy went, hogs followed. The agriculture of the Corn Belt was essentially defined as raising corn to feed livestock, and while cattle and chickens also eat corn, as the region was settled, pigs were the main consumer, as fattening hogs was the easiest way to get corn to market. In fact, pigs were often referred to as "cornfields on legs." Pigs remained popular for the same reasons they were popular since prehistory: prolific, tasty, eat anything, easy to preserve.

But pork outgrew its down-on-the-farm image early on, coming alongside corn to transform the Midwest into an industrial powerhouse. Massive hog drives predated the brief age of cowboys. Carl Sandburg called Chicago "Hog Butcher for the World," though Cincinnati was our first "Porkopolis." The big meatpacking companies grew up along with the Chicago Union Stock Yards, but so did a wide range of allied businesses. Even today, with so much changed, no one else in the world produces more pork than the American Midwest.

For most of the country's history, pork was the almost inevitable meat at any meal.[5] However, in 1909, for the first time, beef production surpassed pork production in the United States.[6] Pork and beef traded the top spot regularly for the next four decades, but in 1953, beef pulled decisively into the lead.[7] For a little more than half a century, pork lagged behind beef in the United States—though that still represented a lot of pork (more than twenty-three billion pounds of pork produced by U.S. meat companies in 2013—only about 10 percent less than the amount of beef processed[8]). Then, in 2015, due to rising beef prices, as well as the increasing popularity of pork products, pork again passed beef.[9]

There is an interesting parallel between pigs and the corn that made their abundance in the Midwest possible. No other plant is better at capturing and storing the sun's energy than corn,[10] and no other animal is better at turning energy into meat and fat than a pig.[11] They are well matched.

Hence, though the story begins long ago and far away, it will follow the spread of pigs around the world and then move with them across the growing United States and into the American Midwest.

What's Ahead?

The first chapter will introduce the pig, as the story that follows often hinges on the behaviors, peculiarities, and qualities of the animal. Then, the historic narrative starts out about twelve thousand years ago, in the vast region that witnessed the slow and sometimes ill-defined conversion of wild boars into pigs. The tale moves on to the impact of civilization and growing cities, as well as the changing relationships between pigs and humans. The discussion will then move quickly through the millennia and on to the introduction of pigs to new lands during the Age of Exploration. We'll follow pigs across the early, evolving United States, and then will settle into exploring their place in the Corn Belt. Next, we'll look more closely at how pigs translate to cuisine. Then, the focus will be more on the present, though with consideration of its connection to the past. We'll meet some of the people who work with pigs, from farm to kitchen, and consider the current explosion in pork popularity. The final chapter discusses some of problems associated with pigs, how the problems are being addressed, and why pigs are still viewed as a solution to other problems.

To gather this information, I have read a lot of research reports, history books, and scientific papers, but I have also visited farms and food processors, museums and historic sites, kitchens, competitions, and more, and I interviewed people who know pigs, work with pigs, study pigs (past and present), raise pigs, process pigs, cook pigs, and even run pork-related events, to get their insights and to help reveal how important pigs are, not only economically, but also historically and culturally. (Names and titles of those interviewed are listed in the bibliography and sources section.)

The goal of the book is not to promote the consumption of pork, but rather to underscore the impact of food and agriculture on world events. Pigs and hogs have been a significant part of human and culinary history. They were the first food animals domesticated and they are the number one meat in the world today. As developing countries have become wealthier in recent decades, the demand for meat has increased dramatically and will continue to escalate, putting food animals in the news as well as on the plate.[12] Even for those who don't eat pork, it's worth knowing a little about pigs' impact on cultures, history, and the economy—because their impact is wider than just triggering an increasing number of baconfests. History would have been quite different without *Sus scrofa domesticus*—and we can only guess what impact they'll have on the future.

CHAPTER ONE

Meet the Pig

Before addressing where pigs started and how they got to where they are today, a few details on what pigs are like and what they need will be useful. But first, though it may seem obvious, I'll define what "pig" means, both in this book and elsewhere. I'll also relate some of the other words people use when talking about these creatures, as well as a bit about where they fit in the animal kingdom.

What's in a Name?

Pigs are *ungulates*, mammals with hooves, and *artiodactyls*, even-toed ungulates. Other familiar artiodactyls are the deer, camel, giraffe, sheep, cow, and goat. Pigs are oddballs in this group, as most artiodactyls are herbivores, but pigs and their close kin are omnivores. They eat everything.

Pigs are *suids*—mammals in the family *Suidae*, which, in addition to the creatures we know as wild boars and pigs, includes such porcine cousins as the warthog, peccary, babirusa, and javelina. The most widely known yell for calling pigs—Soooie—is most likely derived from the family name, *Suidae*.

Under the family heading, we have the genus *Sus*, which is occasionally used as a generic term in discussions of early pig history when it is unknown if one is speaking of wild boars or domesticated pigs—because both fall into the genus *Sus*, with the wild boar species being *Sus scrofa* and the domesticated pig being *Sus scrofa domesticus*.

Swine refers to all the short-legged, omnivorous mammals in the family *Suidae*, though it is most commonly used to refer to domesticated pigs.

The word *pig* can be used to refer to any member of the family *Suidae*, but it is most commonly applied to *Sus scrofa domesticus*, to differentiate it from a wild boar. That is how I will be using it, unless something more specific is required.

Technically speaking, however, *pig* means "a young swine." It becomes a hog when it is mature, which, depending on the location and source one consults, can refer either to sexual maturity or a weight topping 120–150 pounds.

Before it is weaned, a pig is a *piglet* or a *suckling pig*. Before a female pig produces her first litter, she is called a *gilt*. After the first litter, she is a *sow*. A *boar* is a male pig that has not been castrated. A *barrow* is a male that has been castrated. There are more terms than these, but these are the ones that most commonly occur in discussions of pig raising or research.

What They're Like and What They Need

It is almost impossible to separate what pigs are like from what they need, since the latter largely defines the former. The natural range for pigs, as well as for wild boars, is determined by the creatures' intolerance to heat. Neither wild nor domesticated *Sus* species have sweat glands, so they cannot moderate body temperature. As a result, their habitats tend to be either cooler climates or, if in hotter locations, in areas with abundant shade and water or mud in which they can wallow. So, in warmer climates, they will spend the heat of the day in shade, water, or both. Without shade, a place to wallow, and plenty of drinking water, the animals, both domestic and wild, will quickly die if the temperature gets above ninety-five degrees Fahrenheit.[1]

Pigs also sunburn quite easily. Shade and mud are the traditional solutions to this problem. Today, since it's illegal for commercial farmers to use sunscreen on their pigs (it affects the quality of the meat), and having every farm near an adequately large forest is not an even remote possibility, shelters must be supplied by those who raise pigs.[2]

Water is vital for another reason. Pigs drink a lot—up to fourteen gallons a day.[3] Sleep is another major priority of pig life. Pigs actually spend about half of every twenty-four-hour period asleep. A good place to take a nap, somewhere both safe and reliably available, is one of the three elements that help create an ideal territory for a pig in the wild, the other two elements being a place to scratch and a place to defecate.[4]

When not sleeping, pigs tend to wake up only as much as needed to drowsily pursue food and water. Pigs consume about six pounds of feed per day, and considerably more if the pig in question is a lactating sow (about a pound and a half more per piglet, up to twenty-four pounds of food per day).[5] Pigs are impressively equipped to eat, with powerful jaw muscles. In fact, most of the facial muscles of all suids focus on opening or closing the mouth, generally with remarkable strength and speed.[6] On top of that, they have teeth suitable for an omnivore, with sharp canines and incisors for cutting and tearing, and molars designed for crushing food. Their teeth are, in fact, similar to those of a bear.[7]

While omnivores by definition eat everything, for most animals, the term means they eat both plants and flesh, but have some limits or preferences. Wild boars and pigs, on the other hand, really do eat everything: grass, roots, nuts, fruits, tubers, snakes, lizards, earthworms, frogs, eggs, and young or infirm animals. They'll also eat garbage, carrion, and even excrement, especially the cellulose-rich variety left behind by grazing herbivores. In seaside settings, they consume fish, mollusks, crabs, and other vertebrates and invertebrates that wash up on or wander along the shore. Their powerful jaws make even coconuts accessible.[8]

Of course, this extreme diet generally only applies today to pigs in the wild. Pigs raised on modern farms are more likely to be eating corn, soy, vitamins, and other carefully controlled foods, to ensure the safety of the meat and the consistency of flavor. (A pig's flesh takes on flavors from its diet, for good or ill. Cooking pork raised on fish, for example, is not a pleasant experience.) In addition, bringing pigs inside and controlling food also reduces risk of trichinosis, an infection caused by a parasitic roundworm, which pigs (and most other carnivores) pick up when they eat infected animals.[9]

Not all domestic pigs are eating completely controlled diets. People who raise heritage breeds for the gourmet market generally still encourage a degree of foraging, at least in the autumn, when mast (nuts that accumulate on the forest floor) is abundant. Acorns and beechnuts are historic favorites that are very popular for fattening upscale pigs. There are also cultures that still expect pigs to scavenge for their food, whether foraging in forests or eating garbage and keeping things tidy around human habitation.[10] But in the West, even the feeding of table scraps to pigs is often illegal, since it can lead to the spread of disease, because pigs are susceptible to, and can in turn pass along, maladies that affect humans (such as influenza).[11]

Controlled feeding doesn't mean pigs aren't still helping us "clean up." Malcolm DeKryger, president of Belstra Milling Company, an Indiana feed company that also operates six pig farms, notes, "Pigs are unique in that they can eat corn in many forms, either straight from the field or the protein-rich by-products of baking or making ethanol, which is great for us raising pigs, but it is also great for the environment, because all these by-products would otherwise go to landfill."

As off-putting as the wild diet might sound, and as uninteresting as the controlled diet might appear, both meet the rather considerable nutritional needs of pigs. Not being ruminants, pigs can't get much out of grasses and other green plants. Protein intake is a critical factor in pig nutrition, especially for the young or for sows during pregnancy and lactation. In fact, if pregnant pigs don't get enough good-quality protein, with the nitrogen it supplies, they may reabsorb their fetuses. However, they don't do well on protein alone, so foraged roots and plants or farmer-supplied grains are also necessary.[12]

As for pigs getting bored with farm food, DeKryger, who has a master's degree in monogastric nutrition (that is, nutrition for animals with single-compartment stomachs, such as pigs, versus cows and sheep, which are polygastric, having four distinct parts to their stomachs), explains that "all the pigs care about is that they get the same quality and taste in the same amounts, plus as much clean water as they want."

After teeth and jaw muscles, the pig's greatest asset when searching for food is its tough, strong, mobile snout. It is this snout that allows the pig to dig up roots, tubers, grubs, and worms. Both wild boars and pigs allowed to forage can have a tremendous impact on the areas where they root for food. It can be both beneficial, in cases where earth is aerated and destructive insects are eaten, or horrifying, if they are rooting in a cultivated field. In the wild, flora in areas that have been rooted by swine undergoes changes comparable to that created by human settlement.[13]

The pig's drive to root, whether for food or simply to explore, has historically created problems for farmers and ranchers. Pigs can easily destroy a pasture, making it useless for other grazing animals. They will root up fields and crops, and even young trees, and their snouts are strong enough to uproot fences built to keep them out. Eventually, people began containing the pigs, rather than the crops and pastures, because it was easier to keep pigs fenced in, at least if they were well fed, than it was to fence off everything else. For thousands of years, however, the most common way to protect farms, gardens, and pastures has been to put rings in pigs' noses—metal loops that clip on rather than pierce the snout. While the rings still allow foraging and even rooting through leaves for free-ranging pigs, they make it hard for pigs to do any destructive rooting, as pushing against the ring is uncomfortable.[14]

Destruction is, however, one of the things pigs need in order to be happy. It could be something as simple as a pile of straw or an old phone book, but they need to have something to chew up and destroy in order to feel satisfied. Pigs will find something to chew and destroy if it isn't given to them. Unfortunately, that something is often the tails of other pigs. However, pigs can and will get into anything they can find access to—and will try to dismantle it.[15]

Dr. Temple Grandin, who is revered for her understanding of animals, as well as for the handling systems and training programs she has developed to keep them healthy, happy, and unstressed, notes in her writing that pigs are complex. They are sociable but will always fight with new pigs. They can be affectionate but can also be vicious, and they are nasty fighters. Part of Grandin's work has focused on how to accommodate the more charming tendencies of pigs while minimizing the damaging ones. She relates that keeping pigs in large groups is safer than keeping them in small groups. That's because in large herds, there is less aggression, less fighting for dominance than there is in smaller groups. Keeping a few pigs can work if pigs are kept with pigs they know, especially if they are littermates. Grandin writes that the worst thing a person can do when raising pigs is to have a small group and introduce unfamiliar animals; the pigs will attack newcomers and can inflict serious injuries. And pigs need to be socialized, both to other animals that might occupy the barnyard and to the humans who handle them. Handlers may even need to establish dominance, as pigs will attack and bite them, as well. Pigs are smart, but they are very much driven by instinct.[16]

Pigs need novelty but get stressed by change. Stress may not seem like a big deal, except that it has repercussions in pigs. Stress can affect immunity, metabolism, and the endocrine system. This is a benefit for archaeologists, because they can tell from the teeth found near dig sites whether the animals were hunted or domesticated, as domestication (that is, proximity to humans) causes a degree of stress in pigs. However, for pig farmers, there are no benefits. Stress can cause panic or aggression, and it damages the quality of the meat. And a sow that is afraid of humans will have fewer piglets. So, there are important reasons to try to understand what causes stress in pigs and what relieves it, and much research is focused on discovering how to keep pigs healthy and unstressed.

Fortunately, there are some easy things that can reduce stress. For example, pigs need to see humans often to not be afraid of humans. Having something to chew or explore reduces stress. Some stress is unavoidable, such as coming into contact with new pigs or being moved to another area. And because pigs are by nature curious, they are likely to wander into stressful situations on their own (though they appear to be less stressed when they find novelty themselves, versus having it sprung on them by someone else). So, understanding stress and how to avoid it are important factors in understanding pigs, how they behave, how they react, and why raising them is not always easy.[17]

Because of their size, appetites, curiosity, and rooting, all pigs are potentially destructive. However, boars can be truly dangerous—and this refers to domestic males, not wild boars (which are also dangerous, but most people expect that of wild boars). One needs to show male hogs considerable respect. They have sharp tusks and are territorial, and two adult males cannot be kept together. They will fight, and no fight ends without severe injuries. If sows are nearby, fences must be especially strong, to keep a boar from simply smashing through them.[18] (Mating is serious business for male suids. A wild boar during mating season will stay so busy that it can lose up to 20 percent of its body weight.[19])

Movies and children's storybooks have introduced the idea of cute, little piglets that can be cuddly playmates. The little ones are cute. They are curious, friendly, and can even be quite playful. However, one would do well to remember that in a short while, that little bundle of pink sweetness will transform into a hog weighing two hundred pounds or more (and still growing). Plus, little piglets bite, and they have needle-sharp teeth. Their teeth are so sharp, in fact, that they are often clipped, to keep them from injuring the sow when nursing (they can really tear up her udder) or hurting other piglets when they fight for milk.[20] The firstborn gets the front teat, which has the most milk, but all piglets have favorite teats, and they will fight to keep them.[21] Survival is not guaranteed for the young ones, as sows sometimes roll over on their piglets, and may even attack them, though most sows are fairly protective of newborns.

Unless they are hand-raised pets, full-grown pigs are not particularly good with children, especially if there are large numbers of pigs. Even if it's unintentional, larger pigs can knock children down and trample them. While it is not a usual occurrence, hungry hogs have been known to kill and eat children. Remember, small and young animals were among the things pigs consider food options. Pigs will even at times eat their own young. Pigs need not be feared, under normal circumstances and with proper handling, but they need to be respected, and one must keep in mind that they are pigs.[22] Some have been bred for a more docile nature, and many, through careful nurturing and socializing, are amiable. However, if one does not know about a specific pig or group of pigs—how they were bred, how they were raised, or how they react to humans or other stress factors—it is best to use caution.

Initially, domestic pigs developed different traits based on climate and habitat. Pigs were larger in northern Europe, smaller and darker near the Mediterranean, and fatter in Asia. For thousands of years, people were happy to let pigs develop as the environment dictated.[23] Things began changing in the late 1700s. That's when it occurred to people that they might be able to arrange for something better than simply a not-quite-wild pig. More than five hundred breeds and varieties of pigs have been developed over the two centuries since humans started actively working to raise pigs to meet specific needs, whether for meat, for lard, for personality, for size, or for whatever other characteristics might be deemed desirable. With this much work going into breeding, people also began to keep records of the various qualities and features of different breeds.[24]

Breeding pigs has always proceeded faster than breeding other animals simply because pigs are such astonishingly swift and abundant breeders. (As noted in the introduction, only rabbits reproduce faster.) Being prolific has, at times, been a drawback as much as a blessing. When raising pigs, one soon has a lot of pigs. A gilt is sexually mature at five to six months of age. After a gestation period of only four months, the sow farrows, or gives birth to a litter. An average litter is ten piglets, but it can be as many as thirty, and a sow often has two litters a year. By the time a second litter comes along, the previous litter is ready to breed. So, within a couple of years, a few pigs can turn into hundreds.[25]

However, the speed of multiplication is not the only astonishing thing about pigs. More remarkable is the fact that, after more than ten thousand years of domestication, pigs still carry all the genetic material needed to revert to being wild boars.[26] They can revert so quickly that it almost seems to suggest that they have never truly been domesticated.[27] Most people know that pigs readily adopt wild living when given the chance, with offspring reverting to the bristly coat and even bristlier personalities of their progenitors. However, more surprisingly, if one simply feeds a young domestic pig less than it has come to expect, regardless of the breed and even while on the farm, it will grow up with a longer, straighter, narrower head than its parents, and will look more like its wild boar ancestors, ready to take on whatever challenges the wilderness presents.[28] It's almost as if pigs are simply tolerating humans as long as the pigs are fed well.

Because they are prolific, they do tend to find themselves naturally in large groups. While most people now say *herds* when referring to swine collectives, that term primarily applies to domestic pigs, while large groups of feral pigs and wild boars are generally known as *sounders*. This is because suids, both wild and domestic, use a range of small noises or sounds to stay connected with others in their groups.[29] In addition to simply staying in touch, these sounds help pigs communicate everything from being happy to being threatened. They emit sharp barks when startled, soft snuffling noises when they're happy to see someone or something they recognize, and squealing when dinner is on its way. Other sounds might signify being in heat, an intention to attack, checking up on the piglets, or feeling contented.[30]

Much writing about pigs these days focuses on the creatures' intelligence. It may be because people are trying to counter the tendency to interpret pig behavior—a life that consists primarily of sleeping and eating—as stupidity. Anyone who has hunted (or read about hunting) wild boars knows how cunning they can be. Domestic pigs may have lost some of the edge of their wild predecessors, but they are still relatively intelligent. They can be trained to do a number of tasks and tricks, and they are very clever indeed about seeking out whatever they need (especially food). However, their intelligence is not necessarily the thing that sets them apart from other creatures. The study of animal cognition is a relatively new area of research. While lists of "Smartest Animals" are popular, even what constitutes intelligence in animals is still being discussed. However, pigs are never in the top spot, with chimpanzees, bottlenose dolphins, dogs (especially border collies), whales, gorillas, elephants, crows, and parrots generally coming in ahead of them, and cows, rats, octopi, and squirrels making some lists.[31] In many cases, cognition in other animals hasn't been studied, so comparisons become impossible. Pigs do have excellent memories, though that can be a liability, as once something frightens or displeases them, it can be a long time before they unlearn that fear or displeasure. Pigs are more curious than sheep but are not more trainable than horses.[32] More to the point, their intelligence has little to do with their history. So, while the fascinating, historic, tasty, and even controversial aspects of *Sus scrofa domesticus* will be examined at length, their intelligence will not be a theme in this book.

Those are a few of the most important quirks and characteristics of pigs. Some of the tendencies and needs had an impact on where and how pigs were domesticated. Others affected how they were and are kept. All are important to know for those who raise pigs, but most are keys to understanding how pigs behave and how they have interacted with humans throughout history.

And now, on to that long and remarkable history.

SECTION I
FROM THERE TO HERE

CHAPTER TWO

Early Pig

Domestication and Early Civilization

Pigs and humans have a long history together. Current archaeological evidence suggests that only dogs have associated with humans longer than pigs have. In fact, by the time calendars were switching from B.C. to A.D., humans across the Old World already had about ten thousand years' experience with swine. Domestication was the big news for the earlier millennia, followed by learning to live with the creatures, as humans settled down and civilization began to take shape (with settling down and civilization generally only occurring in places where plants and animals were domesticated[1]).

Traditions that in some places still exist began to take shape during the early generations of association, from the Assyrians creating black pudding to the Babylonians using pigs to hunt for truffles. But let's start with domestication.

Domestication: Long Ago and All Over

A discussion of the domestication of pigs offers a broad canvas and lots of blurry lines. There is no precise where or when, and dates keep getting pushed back as new evidence is unearthed. Even what might have initially constituted domestication is a bit indistinct, which makes writing about the early history of pigs a little vague. However, it also makes that history remarkable.

Two things are clear: pigs were domesticated a really long time ago, and all the creatures we think of as pigs are descended from the Eurasian wild boar, *Sus scrofa*.

Of all the animals that were eventually domesticated, only the wolf made its home across a greater geographic area than the wild boar, which, twelve thousand years ago, ranged across most of Europe and Asia, as well as parts of North Africa. However, though it shares much in common with the wolf, the wild boar is more adaptable, more opportunistic, and a less fastidious omnivore than the wolf, which led to early and often-complex relationships with humans even prior to domestication.

The progenitors of all other livestock species tended to be relatively localized, making it possible to identify fairly specific regions where domestication occurred, but with the wild boar, domestication appears to have occurred almost everywhere across their tremendous range. Archaeologists working in widely diverse areas in Asia, Europe, and the Middle East have all discovered evidence of early domestication, primarily large numbers of boar teeth that show signs of dietary changes—the kinds of changes associated with eating food provided by humans.[2]

Just as a specific place where wild boars were first domesticated is impossible to identify, precisely *when* they were truly domesticated is also vague and actually rather difficult to define—not because the evidence fails to suggest time periods when humans were interacting with *Sus scrofa*, but rather because what might constitute "truly domesticated" is hard to nail down with pigs. Wild boars began to follow hunters, having learned that most hunters wouldn't be able to eat or transport everything they killed, so the leavings would provide a tasty meal for the wild boars. When people began to settle down, wild boars hung out around settlements and villages for the same reason: eating garbage and leftovers was easier than hunting, plus humans offered a degree of safety from larger carnivores. People became intentional in their feeding of the wild boars, because having them nearby had distinct advantages. (There were drawbacks too, as they were ill-tempered and possessed razor-sharp teeth and dangerous tusks.)[3]

Generally speaking, domestication means selectively breeding an animal in captivity for use by humans, with the result that the animal differs from its wild ancestors. This definition works for most animals, but it is not always perfectly applicable to wild boars and pigs. As wild boars became more dependent on humans for food, they began to change. Legs and heads became shorter. Bodies became rounder. Even before captivity and selective breeding, they began to be pigs. However, boundaries between wild and domestic remained fairly fluid. Wild boars still hung out around settlements and were regularly fed—something known as "managed wildlife," versus domestication. Domesticated pigs were allowed to forage in the woods, where wild boars often mated with domesticated sows. And some pigs became feral.[4] However, changes did become increasingly evident.

Dr. Melinda Zeder, senior scientist and curator of Old World archaeology for the Smithsonian Institution, has been focusing on this transition period for the last fifteen years of her research. Zeder shares,

> Boars were being managed wild by about 11,700 years ago (about 9700 B.C.). Ice and cold weather were still retreating, but in the Near East and Fertile Crescent, the climate had become warmer and wetter—even warmer and wetter than today. Forests grew. People began settling into little economies within resource-rich areas. People still hunted, but they were relying on a combination of wild and managed animals. By 11,500 years ago, people were putting young, loosely managed wild boars on boats and taking them to islands. Then, from 11,000 to 10,500 years ago, there was a gradual increase of control over animals, up to domestication. By 8500 B.C., pigs were definitely under human control. Evidence suggests that goats and sheep were beginning to be domesticated at this period, as well, but there were a lot of pigs by this time.

Archaeologist Dr. Max Price, whose doctoral research focused on pigs and society in this era of developing civilizations, explains what researchers look for when examining skeletal remains to determine the stage of domestication. Price relates,

> I look for clues to reconstruct the life cycle of the pigs. For example, a feature on the teeth called linear enamel hypoplasia is caused by an interruption in enamel growth due to stress. Hypoplasias tend to increase when pigs are penned. We're particularly interested in sites where early agriculturalists began penning pigs, such as Domuztepe, a sixth millennium B.C. site in what is now southern Turkey. The name Domuztepe actually means "pig hill." What I see in samples from these sites is a huge spike in the incidence of linear enamel hypoplasia. What's interesting about the Domuztepe site is that we've found starch grains in the tartar left on the pig teeth. This means the Domuztepe pigs were being fed cooked cereal grains—good evidence for some of the earliest penning of pigs in the Middle East.

While pigs were domesticated independently in many places, evidence currently suggests that the earliest domestication may have been in the Near East. The village of Hallan Cemi, in modern-day Turkey, was occupied around eleven thousand years ago, and archaeologists have found evidence that the residents of Hallan Cemi had domesticated pigs—even before anyone in the area was practicing agriculture.[5] A great deal of research was done after the site was discovered in 1989, and tons of material was salvaged from the site before it was flooded by the construction of a hydroelectric dam.[6] Dr. Zeder relates,

> We recovered two metric tons of animal bones from the Hallan Cemi site. In studying these bones, we've been seeing some interesting patterns. I have been analyzing the bones for 15 years, to construct profiles of ages of pigs, to determine if populations were hunted, managed, domestic, male or female. Hunters go after bigger animals, but herders want to keep a herd viable, so there are different killing patterns. Plus, you don't need a lot of males around when herding; you want mostly sows. The tendency was to kill pigs at six months of age, because they grow so fast. Did this help them manage large, rapidly growing herds? Research often leaves us with hypotheses to be tested, rather than answers.

Archaeologists continue to find more and more locations where there is evidence of humans and wild boars interacting and domestic pigs appearing. Research at sites in Turkey, Iraq, Iran, Syria, Egypt, Switzerland, Sweden, Denmark, the Netherlands, England, France, Germany, Poland, Greece, Turkestan, China, Japan, and Indonesia has already demonstrated the extent of early association of humans and *Sus scrofa*.[7] Most suspect that as more sites are excavated, more locations will be identified.

As blurred and vacillating as the process of domestication was, by the third millennium B.C., pigs were well established pretty much everywhere across Europe and Asia. So were sophisticated cultures that were leaving behind more than just boars' teeth as evidence of reliance on pigs. Pottery began showing images of pigs. Early writings provided an even clearer picture of cuisine, agriculture, and even economics. For example, by 3486 B.C., decrees were being issued by a ruler in China who wanted people to breed pigs both for consumption and for trade with other countries.[8] The more concentrated populations became, the more they needed easy access to food—and nothing was easier than pigs.

Cities and Civilizations

Though the domestication of pigs may have slightly predated agriculture,[9] it didn't do so by much, at least in most places. People would not have been able to settle without a nearby food supply, and while wild plants could support smaller villages, agriculture was needed once cities began to develop. And with the advent of larger settlements and the need for more protein, pig numbers increased, as well.[10] (Other animals had been domesticated by this point, but because oxen could pull plows and sheep offered wool, people were generally hesitant to kill them for food.)

Small villages, nomadic herders, and hunter-gatherers still existed (and still do exist), but across Eurasia, people began gathering into larger settlements. In the Jordan Valley, for example, Jericho had a population of three thousand by around 8000 B.C. Camps in the area show that hunters in the Jordan Valley had been butchering boars nearby since before 10,000 B.C., but smaller skulls among the remains found inside the city demonstrate both that boars had changed into pigs and pigs had moved in with humans.[11]

In China, evidence shows that pigs were domesticated by around 8000 B.C.[12] Though game, especially deer, remained popular, by around 6000 B.C., pigs were not only domesticated but were part of an increasingly sophisticated agricultural society. While grains tended to be divided between millet in the north and rice in the south, pigs were in evidence in both regions.[13] Dr. Price notes, "In China, pigs were penned and intensively managed much earlier than they were in Europe. This might be why, when Europeans eventually got interested in targeted breeding of pigs several thousand years later, they turned to Chinese breeds for their adaptability to more intensive penning."

Pigs would grow in importance in China until pork surpassed all other meats combined. The impact of pig domestication on Chinese culture is underscored by the fact that the Chinese ideograph for *home* consists of the character for *swine* beneath the character for *roof*. This "home is where your hog is" iconography reflects how most pigs were raised in China. China was a land of villages, and most pigs were raised by families for consumption by the family or at most by the village (a practice that would persist until recently, despite other major changes). By 4300 B.C., pig production had grown to the point where politicians began to think it would benefit from government oversight.[14] That said, Chinese governments in 4300 B.C. were local, as China was not actually unified until around 221 B.C. But government interest underscores that pigs had become key aspects of life.

In western Asia, the region known as the Fertile Crescent would give rise to some of the world's earliest civilizations—not mere settlements or villages, but cities and groups of cities with complex cultures, politics, art, and written language. Sweeping in a rough arc from the Nile Delta and Mediterranean Sea to the Persian Gulf, the Fertile Crescent had a mild, wet climate at the time these civilizations first emerged, making the region well suited for agriculture. In addition to raising crops, the people in this region depended on four domestic ungulates for their meat: sheep, goats, cattle, and pigs. The remains of pigs still show the presence of an occasional wild specimen, but agriculture and raising domesticated animals had become the norm by the time these civilizations came along.[15]

The city of Sumer, located in the part of the Fertile Crescent known as Mesopotamia, between the Tigris and Euphrates Rivers, was the location of the world's oldest known civilization. It was settled between 4500 and 4000 B.C. Sumerians created the first vehicles with wheels and the first system of writing: cuneiform. (And the earliest recipes yet discovered were recorded by Sumerians in cuneiform.) While Sumerians kept large herds of domesticated pigs, the pigs were expected to scavenge for most of their food, though this was supplemented with barley, no doubt to keep the pigs at least partly reliant on humans. The scavenging, however, explains the continued presence of wild boars, as both wild and domesticated animals would be attracted to the same food sources.

Not as early but destined to be far larger were the civilizations that arose in the Indus Valley. The earliest Indus Valley cities, such as Rehman Dheri, emerged around 4000 B.C. We know they had pigs because pigs appear in the artwork—carefully rendered motifs of domesticated animals on bowls and other pieces of pottery.[16]

Though settlement started centuries earlier, by 2700 B.C., the Harappan civilization was well established in the area of the Indus Valley now occupied by Pakistan. This great urban civilization would thrive for a thousand years. The two main Harappan cities, Mohenjo Daro and Harappa, had populations of forty thousand and fifty thousand, respectively.[17] Agriculture was sophisticated and varied, with field crops (primarily grains and legumes) and livestock that included goats, sheep, and cattle. But here as elsewhere, pigs were popular. Pig bones, terra cotta pig figures, and pigs drawn on pottery bowls reveal the presence of domestic swine in the culture as well as the diet.[18]

Among the most influential of the Fertile Crescent civilizations was Babylon. It became the capital of southern Mesopotamia around 2000 B.C., and that part of Mesopotamia became known as Babylonia. While Babylon is known for many things, from successful warriors to hanging gardens to amazing architecture, the thing with the most far-reaching impact may have been Hammurabi's famous laws. Law number 8, which includes stealing pigs, prescribes dramatically different consequences for stealing from the court versus stealing from an ordinary citizen. This hints at another development in the world. With the rise of the great city-states, both in Mesopotamia and China, food resources and diet began to differentiate along class lines. As long as societies were simply trying to survive, everyone ate the same thing, with differences only in quantity, as dictated by a group's traditions. But with the rise of rulers, priests, nobles, and warriors as separate classes, the culinary world divided into high cuisine and low. High cuisine tended to be meat-heavy and offered sweets, intoxicants, and other delicacies that would not be available to the lower orders. Of course, even high cuisine was not monolithic, with the palace having more than the nobles.[19]

Dr. Zeder adds,

> In my research in the Middle East, I observed that pig numbers went up when a town was more self-sufficient and declined when an area became part of a larger, more complex culture. Pigs are ideal for self-sufficiency. People could raise them even if they didn't have land. What that led me to is that pigs are not a good resource for large, managed economies but are ideal for urban poor. That makes raising pigs almost a political statement—a way for people to remain independent in a highly controlled economy. The archaeological evidence supports this. In Mesopotamia, we haven't found evidence of a high number of pigs in elite areas, but the number is high in poor areas. Pigs offered a degree of autonomy. Of course, this identification of pigs with the poor probably contributed to increasing contempt for pork among the wealthy.

However, while the rich might have been biased against pigs because of their association with poverty, that didn't mean they wouldn't eat the meat. An old Sumerian saying suggested that "slave girls should get only lean ham, as pork is too good for them." Fat meat, being in short supply, was always prized, so pigs, or at least certain cuts, were valued.[20] Also, by the time Babylon was thriving, people had discovered that pigs were good at finding truffles, so the wealthy had another reason for loving pigs.[21]

On the western edge of the Fertile Crescent, Egyptian civilization emerged in the fourth millennium B.C. Egypt had a well-developed agricultural economy, and pigs were among the several animals raised as food, domesticated from the wild boars that inhabited North Africa. Written records relate a history of large herds of pigs and enjoyment of pork, and pottery statues and votive figurines suggest widespread popularity. In addition, abundant pig bones have been found near villages and other Egyptian sites,[22] so pigs were clearly on the menu.

Assyria, which arose soon after Babylon did, was a dependency of Babylon. To be honest, things were considerably more complicated than civilizations simply arising. Different groups at different times emerged, moved, got into wars, expanded their power, and then dropped back. But after considerable conflict and relocating, Assyria was a major empire in the region, and Assyrians became known for architecture, fighting skills, and cruelty. Nineveh was Assyria's oldest and most populous city, but Tyre was home to butchers famous for their pork preparations. There are historians who credit the Assyrians with the invention of sausage, especially the blood sausage also known as black pudding. Of course, given how widespread pork consumption was, it is possible that any of the other claims of primacy in sausage invention are equally true, just as pigs were domesticated in many places. But it is to the Assyrians that the French attribute their famous *boudin noir*.[23]

Recent archaeological exploration in the British Isles turned up the fact that the builders of Stonehenge, working before 2000 B.C., gathered for large, organized feasts during breaks from construction of the now-famous monument. Observations about how food was distributed have intrigued researchers, who now say evidence does not support the idea of slave labor being used. They also discovered, through the huge numbers of bones left behind, that the vast majority of meat being consumed, either boiled or roasted, was from pigs.[24]

So, pigs were well established among the world's oldest civilizations. While many pigs were still expected to roam fields and forests, searching for food, urban foragers and sty pigs fed table scraps were well established by the third and second millennia B.C. By 1500 B.C., pig raising was well defined and regulated, and the next developments would largely be refinements of the system.

Not Everyone Loves Pigs

Even as pigs became almost universally domesticated and adopted, concerns, biases, and taboos were beginning to emerge. In some cases, the taboos are still observed. There are numerous theories as to the reasons various groups would discourage or even forbid the consumption of pork. Of course, given the time span and the variety of cultures involved, it is unlikely that everyone would have had the same motivation for their preferences or prohibitions. It is more reasonable to assume that multiple theories apply, varying by location, culture, and time period.

One theory is related to the limits of where pigs thrive. Pigs require either cool weather or some way to cool off, such as abundant shade and water or mud in which to wallow. The current climate and terrain in many of the places that forbid, or simply dislike, the consumption of pork fail to meet these requirements, so it has been proposed that maybe the inability to raise healthy pigs would suggest to people that they shouldn't be eaten. This makes a great deal of sense for hot, dry locales.

Cultural, social, or economic biases may have played a role in creating some taboos. They certainly have had an impact on preferences. One theory currently gaining traction is that the Mesopotamian attitude about pigs being for poor people would have made others more inclined to reject pigs as viable food animals. It is not hard to imagine that acceptance or rejection by another group would affect the opinions of others. In Mongolia, I was informed that eating pork is seen as "going Chinese," and since the Chinese are traditional enemies, Mongolians don't eat pork. (And in China, I heard regularly that only barbarians eat the stronger meats, beef and mutton, that the Mongols prefer. So, it goes both ways.) It seems reasonable that similar dynamics would exist elsewhere. Of course, it's also possible that, rather than snobbery, the fact that poor people could remain independent with pigs might make governments disinclined to allow pigs.

However, theories of inappropriate climate, adopted bias, or opinions of neighbors do not account for the food prohibitions found in the Mosaic Law. Pigs are far from being the only creatures banned in Judaism. All bottom feeders, carrion eaters, and scavengers are taboo, as are all animals of prey,[25] along with a wide range of other animals that can transmit diseases to humans. These food prohibitions, along with other Mosaic laws (including washing one's hands after touching the sick and burying excrement well away from where people live), were clearly understood to be health rules, as the promise given in Exodus 15:26 is that if these laws are followed, Israel would suffer none of the diseases that the Egyptians suffered.

I have always considered divine revelation to be the most reasonable explanation for these far-reaching and remarkable health laws. However, I might also concede to the theory of inspired observation. The Egyptians considered dung an antiseptic and put it on wounds (which is why tetanus was a leading cause of death in Egypt), so the ideas were clearly not carried over from Israel's captivity. But the Israelites might have been more attentive to cause and effect. Observation of patterns of the occurrence of disease is, in fact, how most medical discoveries have been made, so it is not hard to imagine it taking place here.[26]

Social and cultural biases and religious taboos will always have an impact on one's actions. However, if disease is the primary concern, in developed countries, modern science and carefully managed feeding protect consumers from disease in commercially raised pigs. Also, pigs are now generally raised far enough from large population areas that the risk of direct transmission of influenza is reduced. Of course, health issues will always be a concern when consuming wild animals, especially carnivores, so make sure you know what the risks are if you fancy a bit of game.[27] But if no tradition, belief, or food allergy precludes consumption of pork, one need not worry, at least with commercially raised pork.

CHAPTER THREE

Old World Pig

Pork and the Rise of Familiar Cultures

In the first millennium B.C., people groups and civilizations began to rise that would give the world ideas and cultural norms that are still with us today, from the Celts inventing soap to the Romans introducing the idea of ending a meal with dessert. By this time, across Asia, Europe, and beyond, pigs were not simply domesticated; they were part of the social order. People still hunted wild boar, but they hunted them because wild boars are dangerous and hunting them was viewed as heroic. It was pigs, however, that increasingly became woven into language, stories, culture, and cuisine. They also became even more widespread, as people were exploring and creating new settlements millennia before the official "Age of Exploration" began with Columbus's sailing.

As with all civilizations, there are overlaps and questions of definition, but during the period of approximately 1000 B.C. to A.D. 500, we see the rise and, in most cases, fall of the Celts, Greeks, Etruscans, and Romans in Europe, and see China coalesce into a country, rather than just a region.

The Middle Ages followed, stretching from roughly A.D. 500 to 1500. During this time, the approach to raising and keeping pigs began to change in many quarters. The biggest changes were generally in Europe, as much of Asia, particularly China, had early on adopted a paradigm that would last until quite recently (and still exists in some areas). Pigs also moved into India as well as across the Pacific during this period.

The cultures were different, and many periods witnessed dramatic variations, though often maintaining specific variations for centuries. However, each period contributed something to the future, and through it all, almost everyone relied to a greater or lesser degree on pork.

The Celts

While, thanks to the *War Commentaries* of Julius Caesar, we most often read about the Celts from the viewpoint of the Roman Empire, the Celtic world actually arose much earlier than Rome did. The Celts emerged sometime in the second millennium B.C., though the oldest archaeological evidence—thousands of well-preserved graves—dates to around 700 B.C. The first Celts were salt miners and traders from near Salzburg in Austria, but they were a fairly mobile group, and by about 100 B.C., Celtic tribes had spread from Asia Minor to Spain and northward to England, Scotland, and Ireland.

The Celts would introduce many remarkable changes in farming as they settled northern and western Europe, including rotating crops, using manure as fertilizer, creating a plow superior to anything Rome had, and developing the first harvesting machine. However, the basis for their fame in Rome was their discovery of what salt did for pork. The most valuable trade goods the Celts had to offer were bacon and ham.[1]

The Greek geographer and historian Strabo wrote of the Celts,

> Food they have in very great quantities, along with milk and flesh of all sorts, but particularly the flesh of hogs, both fresh and salted. Their hogs run wild, and they are of exceptional height, boldness, and swiftness . . . [their] herds of swine are so very large that they supply an abundance of . . . salt-meat, not only to Rome, but to most parts of Italy as well.[2]

As Strabo noted, the pigs the Celts raised were large. Their size, however, was just one of the characteristics that defined the Celtic Landrace, a Landrace being a breed of pig associated with a specific region. While many other Landraces were developed over the centuries, at the time Strabo was writing, there were two in Europe: the large Celtic Landrace found in the north of Europe and the smaller Iberian Landrace found in the south. More importantly to the Celts, the Celtic hog was a long, lean "bacon hog," very meaty and ideal for salting and smoking.[3] (The Iberian hogs carried more fat and were more suited to being eaten fresh.) Climate had a great deal to do with the importance of salt and smoke in the north. Celts needed to slaughter their hogs and start the preservation process before winter set in, because their animals wouldn't be able to forage in the snow, while people in southern areas, such as Spain, could let the pigs forage all year.[4]

Pigs were so important to the Celts that they moved beyond mere sustenance. Different cuts conveyed status, with a leg going to the king, a haunch to the queen, and a thigh to the greatest among the warriors. Even in death, it was assumed that pork would be desirable, and joints of meat and even whole pigs were included among grave goods of aristocrats. Myths and legends arose that involved pigs. Wild boars were engraved on helmets, as protection. Swineherds became heroes in some stories. One Celtic myth had the Tuatha Dé Danann, or "People of the Goddess Danu," introducing pigs to Ireland, which, during the Iron Age, was known as the Island of Pigs.[5]

Those who encountered the Celts regularly commented on their heavy consumption of meat, both wild and domestic. "Their food consists of a little bread and a great deal of meat boiled or roasted on charcoal or on spits," wrote Greek author Posidonius. "They are clean eaters, but with a lion's appetite. They select whole joints in both hands and gnaw bits off, or if a bit is hard to tear away, they slice along it with a dirk which lies to hand."[6] So pretty much what one would expect of fierce warriors. However, it wasn't all massive roast joints. The Celts made beer, and it appears that pork and beer stew was popular. Bacon with leeks was another favored combination. However, whatever the form, most of the "great deal of meat" consumed by Celts was definitely from their favorite animal, the pig.[7]

The Classical World

Among the many things the Classical world contributed to discussions of food were a couple of key words related to pigs. From the Romans and Greeks, we got *porcus*, Latin for pig or tame swine, and *laridum*, Latin for lard, bacon, or cured swine's flesh, a word probably derived from the Greek *larinos*, meaning "fat," and *laros*, meaning "pleasing to the taste." So clearly, in the ancient world, pigs were being eaten for pleasure.

The appearance of pigs in early Greek literature makes their importance clear. In Homer's epics, written around 800 B.C., feasts and family meals consistently feature "fat hogs." In *The Iliad*, fierce fighting men are regularly compared to wild boars. Then, in *The Odyssey*, Circe turns Odysseus's men into pigs, which demonstrates her wickedness but also hints at a conflicted attitude toward pigs. When at long last Odysseus returns home, he finds that his swineherd, Eumaeus, is not only faithful to him but is also still faithfully tending the king's substantial herd. The details Homer shares with us tell us not only that pigs were valued, but that, as with the Celts, those who cared for the pigs were held in high regard. The story also lets us know that, among the rich and powerful, a large, well-larded boar was favored over a tender piglet. In fact, Eumaeus apologizes to Odysseus for serving suckling pigs, calling them "servant's pork."

Actually, it was in Greece that the idea of eating for pleasure first surfaced. While ideas of high food and low food had emerged before this time, it was in Greece that the concept of the rightness of the rich eating very different food from the poor developed. That said, ancient Greece was a collection of Greek-speaking city-states, and cultures were not identical. In the warrior culture of Sparta, pleasure and ease were held in contempt, because they made one soft. Still, even Sparta relied on pigs, with the famous "black soup" unique to them created from pigs' legs and pigs' blood, seasoned with salt and vinegar.

In Athens, however, especially after the fifth century B.C., culture and intellect were valued over fighting ability. During the golden age—when Pericles ruled, the Parthenon was built, and Greece's most famous playwrights were creating dramas still read today—the dining habits of rich and poor became more dramatically different. The rich indulged in meat that had not been part of a ritual sacrifice and drank more wine than water. They sought exquisite delicacies and rare exotica, and considered a pig that had died of overeating extremely desirable. The ability of pigs to find truffles was a bonus that was also appreciated by the food obsessed.

As with most spikes in gourmandizing throughout history, there was a parallel burst of food writing. However, all that is left to us of Greek food literature and cookbooks are titles and extracts quoted in other works, but no recipes, just tantalizing hints of meals and dishes, and claims of inventions ranging from "made dishes" (that is, prepared dishes with multiple ingredients) to small plates.[8]

The life of luxury did not go unopposed. Several Greek philosophers, including Socrates, feared that overindulging, and meat eating in particular, could have negative consequences, from unbalancing the humors to making the brain sluggish to overheating the libido.[9]

While the poor might never get a fat hog for dinner, let alone one that died of overeating, pigs were still important, though consumed primarily in the form of sausage.[10] Sausage was the street food of the day, relied on by workers and single men.

With the rise of Alexander the Great, fine dining would get even more complex. Alexander (356–323 B.C.) was Macedonian, and Macedonian cuisine was as epic as anything in Homer (complete with whole, stuffed, roasted hogs). As he captured the world of his day, Alexander created what became known as Hellenistic culture by combining all the things he liked about Macedonia, Greece, and Persia—and that included food.[11]

Alexander's conquests carried him through Greece and Asia Minor, as far as India and across the Middle East into Egypt. Hellenistic cuisine and culture became the standard across this wide realm and would later be adopted by the Roman Empire.

As Greece evolved, so did life and culture on the Italian Peninsula, though there was much blending of populations, ideas, and foods. The Etruscans appear to have gotten their start in around 700 B.C., but when a king was elected in Rome in 616 B.C. (this was before the republic was formed), it was a Greek who had grown up amid the Etruscans. So definitely an intermingling of influences.

Etruscans were seriously food-focused. They considered food preparation a high calling and did all their cooking to music. They were large-scale breeders of pigs, and the only meat for which evidence has been found at Etruscan sites is pork. Music was not limited to the kitchen, as the Etruscans trained their huge herds of pigs to follow the sound of the trumpet. Pigs were, in fact, trained to recognize a specific trumpet, so that many herds could mingle and then be separated by the call of the horn each herd knew.

The Etruscan king indulged in two lavish meals a day. Because wealthy Greeks had only one lavish meal a day, with a couple of smaller ones, the Romans considered Etruscan dining habits excessive and tended to follow the Greek example. Other than the number of banquets, frescoes in tombs indicate that Etruscan banquets were clearly inspired by Greek practices, except in one key element: Etruscan women were free to attend the feasts.[12]

The Macedonians fancied a pig stuffed with wombs. The Etruscans were renowned for their pig's tripe. So perhaps it was to be expected that Romans would love both pig's tripe and wombs, as well as ears and udders.[13] It is not hard to imagine, however, that these odd passions originally grew out of the necessity to not waste anything—because early in its history, Rome struggled to feed itself at all. There were simply too many people to be fed by the surrounding farms, and the city of Rome was frequently on the brink of starvation. It was this that became the driving force of Rome's expansion—they simply took over all the sources of food in Europe and the Middle East. It became cheaper to import food than to grow it. It took nineteen rural workers (at least ten of them farmers) to feed one citizen of the city, and by the time Rome's population topped a million, many of those rural workers were conquered peoples brought back as slaves.[14]

The extensive transportation network set up to move food (and Roman troops), once in place, meant that, regardless of how remote an outpost was, from Britain to Palestine, anyone with money could eat as well as the upper classes in Rome—at least most of the time. And wherever Roman food was carried, pork in some form would be there, because pork was one of Rome's favorite foods.[15] So much a part of the culture were pigs that, even as Imperial Rome was disintegrating and the masses were being kept at bay with "bread and circuses," pork fat was included in the rations given to the poor.[16]

As in Greece, heroes regularly killed wild boars, but most of the pork in Rome was from domesticated animals. As in other cities, the pigs of the poor roamed the streets, living on refuse. Roman families with a bit more money kept a few pigs, though not enough to eat pork regularly. The wealthy, however, not only kept pigs, they often fed them extravagantly, with force-feeding them dried figs and honey wine a popular way to sweeten the meat and fatten the livers.[17] Plus, bacon and ham were coming in from the Celts, especially those in Gaul (modern-day France).

There were no kitchens in worker housing, so most citizens of Rome bought their food from street merchants. (There was no real middle class, just various levels of wealthy upper class and lower class.) Pork-and-leek sausage was particularly popular. Sausages, in fact, crowd the pages of the recipe collection with which Apicius defined dining in Imperial Rome. In addition, Apicius (whoever he was—no one is really certain, and he may have been more than one person) offers entire chapters for different parts or types of pork (shoulder, loin, sow's belly, liver, kidney, fig-fed pork, bacon, and so on). There is also a chapter for wild boar. That said, meat was a luxury for most Romans.[18]

Roman culinary ideas, many of them reflected in Apicius, had a huge influence on European cooking for centuries after Rome fell—until the middle of the 1600s in most cases—and some of the ideas recorded by Apicius linger to the present, such as the instruction to serve sausage with mustard or pairing pork with apples.[19] In Italy, the liver sausage in Corsica is clearly descended from the Roman version, and the Lucanian sausages found in Apicius are reflected in many of Italy's dried pork sausages.[20]

While Rome was still a Republic, family meals in the upper classes were generally fairly modest, with the family dining together at home. However, as the empire (and corruption) grew, the wealthy became wealthier, and Rome's upper classes became infamous for outrageous banquets. Wild boars, whole pigs, and multiple pig parts were always part of an Imperial Roman power feast. Banquets were sometimes purely for showing off wealth, but in some cases, the wealthy simply felt they deserved a banquet, as when one Roman general, in demanding a fabulous meal despite dining alone, roared, "What, did you not know, then, that today Lucullus dines with Lucullus?"[21]

The extravagance could not be sustained. Too many poor had to be fed out of the public coffers. Decadence, intrigue, and bloodshed eroded the government, and enemies surrounded the empire. Plus, the climate was changing. The end of the Roman Warm Period meant farming was less productive and more limited, as one could not farm as far north. The western Roman Empire was undone. Constantinople, in the eastern Roman Empire, faced problems similar to those in early Rome, with heavily taxed farmers leaving their land for the city, and the city facing starvation. The measures put in place to solve this problem set the stage for practices in the Middle Ages.

The Middle Ages

With the collapse of the western half of the Roman Empire, Europe was in trouble. With no Roman army to protect travelers and no taxes to repair roads, the food that had been traded so freely suddenly vanished. People starved. It has been estimated that Europe's population shrank by half between A.D. 200 and 600.[22] In Constantinople, wanting to help hungry citizens and hoping to stave off food riots, Emperor Theodosius II issued laws that were intended to keep farmers on their farms. A *colonus*, or tenant farmer, became tied to hereditary property, and while he could not lose his land, neither could he leave it. While this practice was not limited to Europe, and a certain amount of being tied to the land had already taken place in the dissolving western Roman Empire, this was a clear early expression of the concept of serfdom that would characterize rural society during the Middle Ages in Europe.[23]

Regardless of the shortcomings of feudalism, at least Europe continued to have farmers. Not too surprisingly, it also continued to have pigs. In fact, throughout the Middle Ages, pork would be virtually the only meat available to Europe's lower class—though the middle and upper classes raised and consumed pigs, too. The average peasant/villager owned at least three pigs, and the vast majority of the population was villagers. Granted, pork and bacon were more abundant on the tables of the manor houses, monasteries, and law courts, but if a peasant was eating meat, it would almost always be from a pig. Even when other options were available, peasants seem to have preferred fatty pork, though quantity and frequency of consumption would vary by location, season, and era. Pigs also provided tallow for soap and candles, lard for cooking and greasing tools and axles, and bristles for brushes. (Peasants also grew crops and usually had wonderful gardens, so there were other things to eat, except in years when crops failed. But on the whole, while everyone experienced hunger at some point, peasants generally did better than merely subsisting.)[24]

However, the Middle Ages did not consist simply of peasants and pigs. It was an amazingly dynamic period. Of course, when a period covers a thousand years, there are going to be changes. The Middle Ages are generally divided into the Early (A.D. 500–950), Central or High (950–1300), and Late (1300–1500) Middle Ages, with life and cultures altering as the years passed. During this millennium, the continent was transformed from the dual personality of warring barbarian tribes versus decadent, crumbling Roman Empire into what was recognizably Europe. To this period we owe everything from the Gothic arch to the clock to the musical scale to the first wearable eyeglasses. Names, events, and ideas that belong to the Middle Ages include Charlemagne, Alfred the Great, Dante, Marco Polo, Leonardo Fibonacci, Taillevent, the signing of the Magna Carta, the founding of Cambridge and Oxford Universities, and the building of many of Europe's most notable cathedrals. So, while it is true that there were wars, plagues, and famines, these did not make up the bulk of this thousand-year stretch.[25]

As for swine, much of the relationship between them and humans carried over from antiquity, from tolerating urban pigs to hunting wild boar. In fact, hunting wild boar would remain important well into the Renaissance. Wild boar became the food of the wealthy or the heroic. "The boar slayeth a man with one stroke, as with a knife," wrote Edward, Duke of York, in the 1400s. The difficulty of capturing a wild boar, with the very real possibility of dying in the attempt, offered the kind of glory normally associated with combat, and this glory was a big draw for many. However, game was also a key source of protein for those who lived in the castles or manor houses, which is why the nobility guarded their private forests so tenaciously. Wild boar was not the only game hunted, but tapestries, paintings, histories, laws, and legends assure us that it was wild boar much of the time, as they were abundant as long as there were big forests.[26]

Forests were vital to pig husbandry among the peasantry as well, at least until the Late Middle Ages. Rural pigs were expected to forage, and in the fall, foraging focused on the acorns, beechnuts, and other tree nuts that covered the floors of forests across Europe. The right of grazing in the forests was known as pannage or Common of Mast. Mast, particularly the acorns, was the ideal food for increasing both flavor and fattiness in the pigs' flesh. So important was this foraging of mast that forests began to be classified by the number of pigs they could support. To protect the forests, a specific pannage season was set, usually two months in autumn. Limits began to be set as well on the number of pigs that could be accommodated by specific forests, but the numbers were still impressive. For example, in Lusshardt Forest in southwest Germany, more than forty-three thousand pigs per year were fattened in the early 1400s—and this was after limits began to be set due to overgrazing. However, by this time, pannage was no longer a common right but required payment to the landowner who faced the problem of seeing his forest stripped of the food that wild boars would also be seeking. "Pannage" then became the term applied to the payment for grazing pigs.

In England, after the Norman Conquest, the French-speaking invaders began to tax everyone—and their pigs. Taxation of pigs led to some dramatic changes in England, including England's first professional pig handlers. The swineherds for a village would collect pigs from households, get them to a place to forage, protect them, try to keep them from destroying crops and property, and return them at night. While the herd belonging to any one family might be relatively small, the combined herds were often huge.[27] Swineherds had to undergo apprenticeships, as handling the large herds required specialized knowledge. During pannage season, pigs would begin to shed signs of domestication, and rounding them up could be dangerous.[28]

All across Europe, as populations grew and soil became exhausted, people began clearing forests for farmland. By the Late Middle Ages, "Forest Laws" were being instituted, to try to slow the rate of deforestation, but there were no longer enough forests to feed the massive numbers of swine that had always relied on them. Laws became stricter, grazing became more limited, and fees became higher. Pigs had never actually been full-time "farm animals," but that was about to change. Villagers could no longer just let their pigs roam the fields and forests to get fat. The outcome of this change was something known as the "cottage pig," a pig penned near one's cottage and fed almost entirely on scraps from the kitchen. For the first time, pigs in Europe were being bred in confinement. This also signaled a change in society, as the rights of the nobility to charge for grazing in their forests vanished with the trees.[29]

At the same time rural pigs were munching on mast, urban pigs were grazing on garbage. Pigs were ubiquitous on city streets throughout the Middle Ages. The working class cherished their pigs, of course, but bourgeois families also often had a few pigs scouring the gutters. They were useful in urban environments, because they helped keep urban waste under control. And feeding pigs this way was free.

There was a down side to the proliferation of urban pigs. These large and growing hogs were constantly in the way of horses, wagons, and pedestrians. They customarily invaded vegetable gardens. They pushed themselves in where they weren't wanted, often upsetting pails or knocking down vendors' stalls (vendors who were likely as not selling pork pies or sausages to workers). The first attempt to solve this problem came in 1131, after a street pig in Paris dove between the legs of a horse ridden by Crown Prince Philip, son of Louis VI. When the horse bolted, the prince was thrown, and he died of a fractured skull. An edict was passed forbidding the raising of pigs in town. The edict was ignored, and four centuries later, Parisians still had their pigs.[30]

Another drawback to urban pigs was their waste. Cows were uncommon on city streets, and horses were not nearly as numerous as pigs, plus both cows and horses produced dung that pigs willingly consumed. People in the Middle Ages generally dug privies and cesspits to contain their own output.[31] So pigs were the primary creators of problem waste. As rain washed the excrement into the river, polluted water became a serious issue.[32] It would be a few centuries before germs were discovered, but the negative effects of pollution were sufficiently noticeable that, by the Late Middle Ages, a considerable amount of urban planning was focused on how to manage urban pigs and their waste.[33]

While cities and feudal estates played important parts in shaping the Middle Ages, the monasteries may have played the biggest role in lifting Europe out of the devastating aftermath of the collapse of the Roman Empire. To grow a civilization, one needs to move beyond subsistence farming. For that, markets are needed. This is where the monasteries stepped in. Now, it's worth remembering that in the Middle Ages, the eldest son inherited both title and property at his father's death. The options open to younger sons were pretty much becoming a soldier or joining the church. Women's options were, generally, marriage or the church. Hence, a lot of well-educated men and women—most of them highly motivated to make the impact that birth order or gender would otherwise have denied them—entered the clergy or joined religious orders. This makes it less surprising that monasteries and convents were so industrious.

Monks threw themselves into researching improved farming techniques and equipment, to help improve yields. They shared knowledge and inventions with others, but they also cleared fields, improved land, and raised their own animals and crops.

More food grown meant more food that needed to be processed. Monasteries began taking over food management and processing around A.D. 900. Individual farmers could not afford to build mills, but monasteries could. They also had the resources and influence to coordinate trade, which they did through weekly markets and annual fairs. This new and growing retail food system made variety, as well as abundance, available. The monks began producing cheese, wine, beer, and bread, and these were sold at the markets, but so were the crops and animals raised by villagers. Pork, being the most easily and deliciously preserved of the meats, appeared in the form of sausages, bacon, and hams, as well as lard, which (unlike butter) has the advantage of staying viable without refrigeration. So the larders (originally, places in the home for storing lard, bacon, and salt pork) for both monks and villagers could be well stocked. And as a bonus, discarded distiller's grain from beer brewing and whey from cheese making could feed more pigs.

Economies of scale really do make things cheaper, as do specialization and centralized processing, so this proved to be a boon for everyone. The abbots collected fees and guaranteed that weights and measures were honest. Villagers had access to more technology than would have been possible on their own. Everyone was making more money and eating better—enough better that the population began to increase. The monasteries had become essentially the agro-industrial corporations of their day.[34]

The reestablished trade routes made "high cuisine"—the food of the wealthy—more widespread. While the working classes were fairly locked into what was local, everyone with money could import whatever other rich people were eating. Soon, the cuisine of the wealthy was fairly similar, regardless of location. Often, the only thing rich and poor had in common was pork. While the peasant ate sausage or salt pork, accompanied by local bread and whatever fruits or vegetables were in season, the wealthy might have a roasted wild boar, in addition to a stuffed pig, but might add to this a peacock, a flamingo, rabbits, quail, a sturgeon, minnows, eels, many sauces, a wide range of pastries, and whatever else might show off wealth and the skill of the head cook.[35]

So, people were eating pretty well—for a while. And they would eat well again, but events conspired against unbroken prosperity. Farming suffered from a combination of exhausted soil and the drawing to a close of the Medieval Warm Period in the late 1200s (Greenland had actually been green during this warmer period) and the decline into the Little Ice Age of the 1300s to the mid-1800s. Once again, farming could not be done as late in the year or as far north. Too much rain in 1314 and 1315 ruined crops and brought famine and starvation. When Genghis Khan and his Mongol army swept in from the northeast, the fleas on the Mongols' horses carried a disease (possibly anthrax) that wiped out huge amounts of livestock (though pigs seemed immune—making them even more important than before). This dramatically worsened the famine. The lack of food weakened Europeans, setting them up for the real devastation to come. Carried along trade routes, the bubonic plague, or Black Death, arrived in 1347 and, by the time it was done, it had killed between 20 and 45 percent of Europeans.

However, plague had the effect of improving life for survivors. With so much of the population gone, there was plenty of food for whoever was left. In addition, with so few workers, wages went up. So, the poor were suddenly not so poor. In fact, before the plague, dining among the aristocracy was almost defined by the consumption of large amounts of pork—up to three pounds a day. After the plague, it was the average worker who was eating three pounds of pork per day—which is likely the cause of the aristocracy running after ever more exotic foods, so they could still have something to which peasants did not have access.[36]

After the Black Death, there was a solid rebound before the Middle Ages officially ended, and much of what would follow actually had its genesis in this period. Many of the ideas that would flower in the Renaissance had their roots in the Late Middle Ages. Names and inventions often associated with the Renaissance, from Petrarch's philosophy to Guttenberg's printing press, actually belong to the Late Middle Ages. It was a long, slow dissolve from one era to the next. However, the Renaissance, with the "Age of Exploration" that it spawned, was about to change a lot more than just Europe.

Asia and Oceania

As mentioned in the previous chapter, pigs were domesticated in China very early on. From the beginning, they were numerous, widespread, and generally penned. However, the China of that earliest era was a region of multiple, usually warring, feudal states. The first time a substantial number of these states were united was in 221 B.C., when Qin Shi Huang established the first great Chinese empire and became China's first emperor. This unification did not include all that makes up China today, and it did not last for long, but it was the beginning of what we'd recognize today as China.[37]

However, though the political composition of China was changing, and would continue to change, China's relationship with pigs hardly varied at all. In fact, in rural areas of China today, pig raising and use is nearly the same as it was when China first emerged as a country. Only in recent decades, as China has become more urbanized, has there been a significant change in how and where most Chinese get their pork.

The Chinese penned their pigs primarily because deforestation and a huge population made it impossible to do otherwise. Chinese pigs were prolific, which was vital with a rapidly growing human population, and they were small enough to bring into the house, if necessary. Most, however, were kept outdoors. Rice hulls were common pig feed, but human waste was a major source of nutrition for pigs. Outhouses were often built connected to pigsties. In fact, the Chinese character for outhouse was the same as the character for pigsty. With a bit of added garbage, a few weeds, and some scraps, one pig could be successfully raised on the waste of one person. Larger families could raise more pigs. Still, the poor had little meat for most of the year, as it took time to raise the one or two pigs they might own. The only certain time of year for feasting, at least for the poor, was at the New Year, when the fattened family pigs were slaughtered.[38]

China became the world's leading consumer of pork—which it still is today. However, the Chinese did not stay at home with their pigs; they expanded across the landscape, settling new places, influencing other people groups, and sometimes simply displacing populations (particularly when they viewed the locals as barbarians). They introduced their language, their culture, their crops, and, of course, their pigs. In other words, they created a much larger China.[39]

China's influence was extensive, helping shape much of the history of East Asia, even beyond its own borders. Displaced people groups pushed from the mainland by Chinese expansion settled Taiwan, Southeast Asia, the Philippines, Indonesia, and New Guinea from roughly 3000 to 1600 B.C. Those who spread out to increasingly distant islands became known as Polynesians. They had populated Samoa, the Cook Islands, and the Marquesas by around A.D. 1. By A.D. 500, they had reached as far as Madagascar, Hawaii, and Easter Island. New Zealand was settled by Polynesians in about A.D. 1000. Polynesians took pigs with them on their travels, introducing them everywhere they went. So, pigs were in Hawaii for a thousand years before the first Europeans saw the islands. Today, it is almost impossible for a tourist in Hawaii to not be exposed to their traditional method of pit-roasting pigs wrapped in taro leaves (another Polynesian introduction).[40]

Northern India was part of the land invaded by Alexander the Great, who carried Hellenism and its dining ideals with him as he conquered. Two years after Alexander's death in 323 B.C., Chandragupta Maurya founded the Mauryan Empire, shifting the center of power in India from the Indus Valley to the Ganges. The Aryans of the Indus Valley, who were the dominant people of the new empire, introduced the Persian-influenced cuisine with which they were familiar, and that included pigs. Wealth and status determined meat consumption, but so did region, with more consumed in the northern part of India, where nomad cultures of Central Asia and invaders from Persia had a greater influence on what was being eaten.[41] However, pigs existed the full length of India, and there were in fact a number of breeds native to India.[42]

Today, pigs make up only about 7 percent of India's growing meat consumption. However, the number of pigs is steadily increasing. Some believe that swine are a potential solution to the problem of feeding India's huge and increasingly meat-hungry population—plus raising and selling these prolific, easily-fed creatures might help lift many out of poverty.[43]

Pigs were part of life in Asia Minor (modern Turkey), Egypt, and much of the Middle East from prehistory and remained so as civilizations grew and changed. They certainly would not have been discouraged when the Celts moved into Asia Minor, and they remained popular when Alexander the Great Hellenized the region—everywhere except Israel. Even Alexander the Great, who tried to force everyone to adopt the lifestyle and cuisine he'd evolved, had left Israel alone. But later Hellenizers were not so kind. Antiochus IV, a Seleucid ruler of the Hellenistic Syrian kingdom and successor to Alexander, was determined that everyone in Asia Minor, the Middle East, and Egypt be as nearly Greek as possible. Most people accepted the new gods and the cuisine that came with them. But not the Jews. Antiochus marched on Jerusalem in 167 B.C., with the intention of suppressing Judaism and making everyone prove their acceptance of Hellenism by eating pork. There were other cultures in the region that ate little pork, so not having pigs had never before truly been a cultural identifier. But when Antiochus and the Hellenizers demanded that everyone eat pork or die, refusing to eat it became a key characteristic people began to associate with Judaism.

The Jews threw off Seleucid rule in 142 B.C., but the area would remain independent only until 63 B.C., when the Romans captured Palestine. However, while the Romans were baffled by the concept that anyone would avoid their favorite meat, they didn't require anyone to eat it.[44]

In Israel today, raising pigs for pork is illegal, other than in a small Arab-Christian region in the north. However, one kibbutz, Kibbutz Lahev, is allowed to raise pigs because most of the pigs are used for medical research. Pigs are useful because they are physiologically similar to humans in many ways. Scientists who would never consider eating pork recognize the benefits of studying pigs. Still, it is an interesting development for a people who were for so long defined by avoiding pigs.[45]

Change on the Horizon

Spain and Portugal spent almost the entire Middle Ages under the control of the Moors, who conquered the Iberian Peninsula in A.D. 711–712. The Moors, being Muslims, did not eat pork. As a result, for the heavily Catholic Spanish and Portuguese, eating pork became a more certain identifier of background than even observing church traditions. Pigs became a political statement. One ate pork to show who one was—and who one was not. As the Spanish Inquisition began (1478), Spain's newly ascendant Catholic rulers used pork to ferret out "enemies of the state," who could be punished if they refused to eat it.[46]

The *Reconquista*, the retaking of the region from the Moors, was completed in January 1492—a year that would become famous for far more than the independence of the Iberian Peninsula. With all the resources and soldiers of the region freed up after centuries of warfare, Spain could focus on other endeavors—such as backing Christopher Columbus in his desire to sail to China.

Columbus had studied the work of Arab astronomers, particularly the work of al-Farghani (Latinized to *Alfraganus* when translated). Columbus recorded in his diary that his own measurements while sailing to Africa confirmed Alfraganus, who had calculated equatorial degrees at 56⅔ miles.[47] While some suggest there was confusion about units of measure (because "mile" had different values in different countries at that time) or possibly that Columbus simply ignored those whose theories put China out of range, he was absolutely certain that he could reach China in less than three thousand miles. The experts of the day disagreed (and were correct in doing so), but Queen Isabella was convinced, and she overruled their rejection of the voyage.[48] Spain wanted Columbus to succeed. A fortune was to be had if they could gain control of the spice trade. And if Columbus was wrong, he'd simply die somewhere out on the endless sea. But he found land, and so certain was he of his theory that he died believing it was Asia, despite considerable evidence to the contrary.

This new Age of Discovery, aka the Age of Exploration, had a stunning impact on the lands Europeans explored, settled, or commandeered worldwide. However, things did not stand still in Europe. The next chapter will focus on where pigs went once Europeans started carrying them along on their voyages, but everything Europe's colonies would experience was anchored, at least in part, in the traditions, developments, and ideas of the Old World. Pigs would travel with the explorers, but concepts of livestock raising, breeding, feeding, ownership, and food preparation would be carried along with the settlers who would follow. Of course, everyone and everything kept going back and forth, and the New World had a huge impact on the Old, as well. But the most dramatic changes were in places that had never before had pigs.

The beginning of the Renaissance closely paralleled the beginning of the Age of Exploration, and it was Renaissance ideas that would fuel the passion for discovery. However, there weren't that many changes in the Old World as far as pigs were concerned. Wild boars were still more prestigious than domestic pigs. While available forests had diminished, thus reducing the number of mast-fed pigs, they hadn't vanished, and urban pigs still roamed the city streets. People still enjoyed pork, ham, bacon, sausages, trotters, blood pudding, and other "everything but the squeal" options. Every European country had multiple pork traditions, and many of those traditions would later become identified with the New World locations the various national groups settled.

One dramatic change did happen in Europe, however. In the late 1700s, it occurred to British landowners that they could create something more suited to tastes, needs, climate, and farming styles than the available pigs, which in the Middle Ages didn't look or act dramatically different from wild boars. With the need to confine the once free-range pigs, farmers looked to China, as Chinese pigs had been in pens for millennia. Chinese pigs had developed swayed backs and shorter legs, and were more docile. They also developed fat in a layer right under the skin, which was ideal in a world still dependent on lard for just about everything that needed cooking or greasing. Chinese pigs were imported, and breeding programs began. By the 1800s, many of the most famous breeds, the ones we now consider heritage breeds, had been developed, and in the late 1800s, breed societies started monitoring breeding practices for these breeds.[49]

English influence is made fairly obvious by the names of many heritage breeds, such as the Berkshire, Hampshire, Suffolk, Lincolnshire, Essex, Leicester, and Yorkshire (also known as the Large White). Breeds were defined by ear carriage (that is, whether the ears stand up straight, known as pricked ears, or fold down over the eyes, known as lop ears), color, face shape, and often by distinctive markings, with these traits occasionally

Photo 3.1. Lop ears and distinctive spots contribute to identifying this sow as a Gloucestershire Old Spot. Photo by Cynthia Clampitt, taken at Jake's Country Meats, Cass County, Michigan. Used with permission.

reflected in their names (such as the Red Wattle or Wessex Saddleback). Breeds were further divided into those raised for meat (known as bacon pigs) and those raised for lard. These pigs would become hugely important, as they provided the breeding stock not only for the increasingly industrialized Old World, but also the stock from which the New World would create its own domain-specific breeds.[50]

Pigs would remain important in the places that valued them from the Classical Period through the Middle Ages. However, in the centuries following 1492, as more and more countries joined the race of exploration, discovery, and acquisition, the story of pigs would expand to new lands.

CHAPTER FOUR

Colonial Pig

Pigs in the Age of Exploration

While many people think of 1492 as the beginning of the Age of Exploration, the goals and attitudes that defined the period, as well as the needed technologies, began developing earlier. Exploration began before then, too.

The desire to travel and learn had long driven explorers, such as Marco Polo in the 1200s and Ibn Battuta in the 1300s, to the world's open roads. Tales of exotic lands fueled interest in travel, but so did the desire for trade. International trade routes had sprawled across the Old World for millennia, carrying spices and silks from East to West, but the tremendous expense added by multiple middlemen along the routes made Europeans wonder if there might be another way to get to China. The capture of Constantinople by Ottoman Turks in 1453 further complicated the spice trade and offered additional motivation to those dreaming of reaching Asia.

Ships that could spend a lot of time at sea were relatively new developments, first appearing in the mid-1400s. Additional masts and sails made oars unnecessary, which made ships more seaworthy. Portugal began charting the northwest coast of Africa in the early to mid-1400s and had traveled as far south as the Cape of Good Hope by 1488. However, it was Spain that would sponsor the trip that would change the world—including how the world eats. The Americas would contribute maize, potatoes, tomatoes, chilies, chocolate, cassava, pineapple, peanuts, and vastly more to the world larder. The Americas, in turn, would gain, among other things, wheat, rice, citrus fruits, onions, European herbs, Asian spices, and, of course, the mixed blessing of livestock.

But first, people had to get where they were going—and feeding people while they were getting there presented special problems.

All at Sea

No matter what the ultimate objective was, reaching anywhere involved a great deal of time at sea. As exploration continued through the 1500s and routes between Europe and the Americas became understood, charted, and familiar, it could still take ten weeks to make the crossing, but at the outset, a voyage might take eight months. The amount of space left on ships once personnel, weapons, and trade goods were loaded didn't allow for the packing of anything more than the very minimum amount of food and water. It was assumed that there would be someplace during the voyage where supplies could be replenished, but this was not always the case. Plus, food that could withstand several months at sea made for a less than balanced diet. In fact, with no vegetables or fruit, scurvy was a major problem on sailing ships for several centuries, causing more deaths than combat.

Starch and meat were the primary foods packed for a voyage. Starch might include dried peas but would most often be ship's biscuit, also known as hardtack or pilot biscuit. This was a rock-hard slab of dough made of flour and water, baked at least twice (which is what *biscuit* means—"twice cooked") to drive out all moisture, and reputed to remain edible for up to fifty years, though it had to be soaked in stew or soup to make it soft enough to eat. Despite its hardness, after weeks at sea, ship's biscuit was often found to be infested with weevils, which made it unappetizing. (Columbus's son, Ferdinand, reported that men waited until nightfall to eat, so they couldn't see the weevil larvae in their bread.)

The meat on these voyages was salt pork or salt beef, though pork was far more common, largely because it was cheaper, price always being a consideration when outfitting a voyage. With no refrigeration at the time, pickled pork and salt pork were what most people were eating on land, but the pork packed for sailors would not generally be the nicer cuts of meat. However, a bigger problem at sea was that there was inadequate fresh water for soaking or cooking the meat, so it was generally far too salty to be pleasant.[1]

Once the earliest European explorers found places they could land, live pigs began to be included on return journeys. Of course, this wasn't necessary for those who headed toward Asia and across the Pacific, but in the West Indies, Mexico, and South America, where Spain and Portugal first landed, protein was in short supply, and pigs were a big part of solving that problem.

The World Divided

Having funded Columbus's voyage, Ferdinand and Isabella viewed the land he found as being theirs. They wanted to make certain archrival Portugal couldn't make any claims, so they petitioned the pope to guarantee their right to the New World. Pope Alexander VI was Spanish born and thus was inclined to side with Ferdinand and Isabella. In 1493, the Pope established a Line of Demarcation, a line that stretched from pole to pole, mostly across ocean, dividing the world between Spain and Portugal. This first line gave Portugal virtually nothing in the New World. When Portugal complained, the line was moved westward. The Treaty of Tordesillas, signed in 1494, confirmed the new Line of Demarcation, which gave Portugal the part of South America that would become Brazil. Spain would be given everything west of the line—so all of the New World, minus Brazil—and Portugal would get everything to the east, excepting countries with Catholic rulers.[2] A number of countries challenged the treaty, especially France and England, but they were ignored.

England did more than simply protest, however. They had their own Genoa-born navigator who believed that Asia could be reached by sailing west. Giovanni Caboto, renamed John Cabot when he moved to England, convinced Henry VII that he could duplicate Columbus's success. In 1497, Cabot succeeded in reaching the eastern shores of Canada and claimed them for England. However, Spain had the most powerful military in the world at the time, so England would not be able to follow up on that land claim for another century.

Columbus remained busy even as the 1494 treaty was being negotiated. On his second voyage, in 1493, he carried seeds and plant cuttings, including the first sugarcane, a crop that would come to dominate the Caribbean. He also transported horses, dogs, pigs, cattle, chickens, sheep, and goats. Clearly, the Spanish were preparing to stay. Of all the animals, pigs adapted most quickly to their new environment. They could find abundant food and there were few, if any, predators. Pig numbers were soon increasing impressively. While other animals were useful, it was pigs that made the biggest difference initially.[3]

Indigenous people had mixed reactions to the introduction of pigs. They were happy to replace their dogs with pigs as sources of protein—and when the Spanish insisted that human flesh be taken off the menu in the Caribbean and Mexico, having pigs available made that easier to accomplish. However, pigs ate everything in sight, including the fruit and roots that made up most of the local diet, plus birds, snakes, and lizards.[4]

What no one could have anticipated was that livestock could be lethal. Pigs and horses introduced the first of the diseases that would devastate Native Americans. The Europeans who brought these animals were those who had themselves survived living with livestock. Of course, no one knew at that time how disease was transmitted. In the 1500s, herds of animals simply meant survival.[5] (But while influenza and other animal-borne diseases were first, the even deadlier smallpox would soon follow.)

The Spanish and Portuguese didn't introduce pigs solely in the places where they settled. They left breeding pairs on any islands they passed, to make sure there would always be food available for anyone who needed it, whether victims of shipwreck or soldiers about to go and claim another part of the region. By April of 1514, Diego Velásquez de Cuéllar could write to King Ferdinand that the twenty-four pigs he had brought with him to Cuba now numbered thirty thousand (though thirty thousand was probably a hyperbole intended to convey the idea of vast multitudes, rather than a precise count).[6]

The conquistadors who came with the pigs were no strangers to pig culture. Three of the best-known commanders of this early era—Hernán Cortés, Francisco Pizarro, and Hernando de Soto—were from Spain's Extremadura, a region known for its pig herds and famous for its *jamón iberico*. The combination of devotion to pork products and the desire to not starve made pigs a major part of Spanish plans of conquest.[7]

Hernán Cortés had pigs with him when he landed in Mexico in 1519. (The large, sometimes dangerous wild boars that thrive in areas along the Gulf of Mexico today are the feral descendants of early Spanish pigs.) The valley of Toluca, in central Mexico, was conquered by 1521 and the city of Toluca was established in 1530. Having been captured and settled by soldiers from Extremadura, Toluca was soon famous for pork sausages and hams in the style of Extremadura.[8]

The introduction of pigs transformed Aztec cuisine, as well as the cuisines of other people groups Spain conquered. Aside from various forms of pork, indigenous people also gained lard and learned about frying food from the Spanish.[9] But the Spanish gained something from the New World that helped them raise pigs: maize—the wonder crop that had transformed the Americas even before Europeans arrived. With this prolific grain of the New World coming in as tribute from conquered natives, the Spanish could take large herds of pigs wherever they wanted to go. In 1531, Pizarro took pigs with him to Peru, and pork soon dominated the meat market in Lima.[10]

In time, explorers and fortune seekers no longer needed to bring pigs from Spain; they could easily get them from Cuba, Hispaniola, Puerto Rico, or Jamaica. It made exploration and conquest all that much easier. When Hernando de Soto landed in 1539 at Tampa Bay, the thirteen pigs he included among his supplies came from his home base of Cuba. De Soto's pigs were all pregnant sows, as the objective was to have a herd large enough to make certain the six hundred men who traveled with him would not run out of food. Despite the fact that some pigs were consumed by de Soto's troops, while even more escaped and became feral or were given to Native Americans to keep the peace, the expedition's little herd had grown to seven hundred pigs by the time de Soto died three years later.[11]

While Spain expanded its conquest of the New World, Portugal worked on claiming its half of the globe. The Portuguese focused first on exploring Africa's coasts, both to access the continent's resources (especially slaves) and to find a way around it. In 1498, Vasco da Gama succeeded in rounding the Cape of Good Hope and sailing to India. He had circumvented the Arab-controlled trade routes, giving Portugal control of the spice trade. The Portuguese would continue eastward, establishing themselves in Indonesia's Spice Islands in 1512 and finally reaching Japan in the mid-1500s.

Portugal claimed and settled Angola and Mozambique on the African continent, transforming the culture and cuisine of both countries. Among the many changes was the introduction of domesticated pigs, adding considerable meat to a diet that had previously consisted largely of dairy and grains.[12]

In 1500, the Portuguese landed in Brazil and began to settle the corner of the New World given them by the treaty with Spain. While most European livestock did not do well in the hot, wet coastal climate of Brazil, pigs loved it. Soon, there were multitudes of pigs, and pork became a major element of the diet for colonists.[13]

In India, the Portuguese made Goa their capital. In addition to new foods, the Portuguese brought the Inquisition with them. Hindus were expected to eat beef, and Muslims, pork, to demonstrate their allegiance to the Catholic Portuguese. While the requirement to eat pork was eventually lifted, today, more pork is eaten in Goa than in most other parts of India, and Portuguese-inspired pork vindaloo remains Goa's best-known dish.[14]

New World silver mines and slavery made Spain wealthy (though, ironically, the overabundance of silver eventually bankrupted the country). Portugal's fortune was made through spices, trade goods in Asia, and slavery in Africa. The two countries would continue to spread their control and influence throughout the 1500s. While a few privateers, such as Francis Drake and Walter Raleigh, would annoy Spain in the mid-1500s, it would not be until after the defeat of the Spanish Armada by the English in 1588 that others would be able to challenge Iberian ownership of the globe. Soon after Spanish control of the seas was broken, other Europeans with a coastline and a sailing heritage—primarily the Dutch, French, Swedes, Finns, Danes, and English—were making claims of their own. The focus of Spain and Portugal began to switch from conquest to protecting their claims.

Other Countries Take to the Seas

While some governments wished to join Spain and Portugal in conquest, many Europeans were thinking of the New World in terms of resources. With Europe's population growing, there wasn't enough land, and forests were dwindling. The Little Ice Age, which started in the early 1300s and continued until the mid-1800s, was at its peak. It had shortened the growing season, so agriculture was crippled in northern climes and, all too often, there wasn't enough food.[15] Plus non-Iberian Europeans were weary of watching Spain and Portugal get rich while they struggled. So, there were plenty of people eager to get in the game. The possible options were either to settle land not yet conquered by Spain or Portugal (most easily accomplished in the New World) or to take things away from Spain or Portugal (generally what happened in Asia).[16]

The Renaissance, with its resurrection of the laws and ideas of ancient Rome, helped establish the mindset with which Europeans now approached the world. By the time colonization began in earnest in the 1600s, these ideas had already been percolating through society for a couple of centuries, and they were not simply accepted, they were ingrained. Among the Roman laws revived, one that had a huge impact on colonization was *res nullius*, which translates roughly as "no one's property." This held that any property for which a specific owner or owners could not be identified actually belonged to no one.[17] It's the concept that guided Romans as they conquered large swaths of "barbarian" Europe. The Roman assumption of their superiority is reflected in the enslaving of millions of people from conquered territories, and slavery was another concept the Renaissance brought back into vogue, after having been essentially ended in Europe during the Middle Ages.[18] And the Roman concept of the inevitability of the success of "superior" races would in time evolve into Darwin's assumption that "the civilized races of man will almost certainly exterminate, and replace, the savage races."[19] Obviously, not everyone agreed with these ideas, but enough did that they shaped much of what happened in the next few centuries.

While philosophy opened the doors, varied motives drove people through them, from avarice to a sense of adventure, conflict to dire poverty, persecution to famine, or simply having nowhere else to go.

The Dutch, who had for some time handled the distribution of spices from Portugal's half of the world, lost that option when Spain and Portugal united (unhappily and temporarily) in 1580. (The Netherlands were fighting a war of independence against Spain at the time, so when Portugal joined Spain, Portugal suddenly became an enemy of the Netherlands.) So, the Dutch decided to find out if they could just go and get their own spices. The Dutch headed first for the East Indies (Indonesia) and then southern India. As the 1600s dawned, the Dutch wrested control of the spice trade from Portugal, though Portugal retained control of Goa in India.[20]

Heading west, the Dutch began settling North America in 1609. It was the Dutch who famously bought Manhattan from the Lenape Indians in 1626 for a chest full of "trinkets" (though in reality, it was trade goods, including cloth and hatchets).[21] It was also the Dutch who, in the 1650s, built a wall along the boundary of New Amsterdam, to separate themselves from the British neighbors who'd arrived. The street that eventually ran along the inside of the wall became, of course, Wall Street.[22]

The Dutch West India Company went after Portuguese trade in the West Indies and Brazil in the 1630s and 1640s, and Dutch colonization of South Africa began in 1652. In Brazil, pigs were well established by the time the Dutch arrived. In South Africa, it's not so clear when pigs were introduced, because the Dutch brought in pigs, but it's likely the Portuguese had dropped off pigs as they rounded the Cape of Good Hope on their way to India and Indonesia.[23]

The French, while also pursuing wealth in the Caribbean/West Indies, began settling in areas that would become Maine and Canada. The French government hoped their explorers could find a Northwest Passage around the continent, so they could reach China and India that way. However, it was also a great region for fishing and fur trading. The first French settlement was organized in 1604, on an island in an inlet of the Bay of Fundy. The French then spread inland, settling a region they called Acadia, an area that includes what are today New Brunswick, Nova Scotia, and Prince Edward Island.

Both pigs and traditional French methods of preparing them arrived with the French settlers. Pigs had definitely reached Quebec by 1609 and may have arrived earlier.[24] The voyageurs, lifeblood of the French fur trade, were so reliant on pork for sustaining their astonishingly rigorous lives that they became known as *mangeurs de lard*, or "pork eaters." The pork meat boiled into soup that fueled the voyageurs was also cherished by hard-working French farmers. Tourtière, a pork pie that is still enjoyed in Quebec, arose in the 1600s. A ragout of pork hocks known as *pattes de cochon* was popular, and yellow pea soup was only one of many dishes that relied on salt pork for flavor and seasoning.[25] That yellow pea soup with salt pork is still so much a part of French-Canadian culture that even today it appears not only in cookbooks, but also in cans on grocery store shelves.[26]

The French were joined in Canada by the Portuguese, English, Irish, and Basque, all in pursuit of cod. However, the salt-pork traditions of these countries became the foundation of the culinary practices of the fishing communities that grew up along the coast. In Newfoundland, for example, salt pork is still a major element of traditional cuisine. It provides the flavor for stews and the fat for frying cod cakes, and scrunchions—crisp-fried diced fat from salt pork—are inevitably served with cod tongues (which are not actually tongues, but white meat from the throat of the fish).

By 1682, French explorer René-Robert Cavelier, Sieur de La Salle, had made his way from Canada, across the Great Lakes and down the Illinois and Mississippi Rivers to a region on the Gulf of Mexico that La Salle named for Louis XIV: Louisiana. Serious colonization by the French began in 1702. In the mid-1700s, settlers from France, the Creoles, were joined in Louisiana by French Canadians who had been driven out of Acadia by the British—a people who became known by a name that is a corruption of Acadian: Cajun. The region developed a cuisine that made heavy use of pigs, with boudin, jambalaya, gratons (cracklings/pork rinds), andouille (a spicy sausage), and tasso (spiced, preserved pork shoulder) becoming iconic.

In 1638, Swedish settlers began building New Sweden in what is now the Delaware Valley. Because at that time Sweden had conquered much of Finland, many of the "Swedish" settlers were, in fact, Finns. The colony of New Sweden spread up the Delaware River into what would become New Jersey. Swedes, too, were great lovers of pork, and their settlements contributed to the growing herds on the North American continent. The colony didn't last for long, as the Dutch, under Governor Peter Stuyvesant, demanded the surrender of all Swedish forts in 1654, but New Sweden did contribute a great deal to early American culture, including the log cabins that seem so inseparable from our image of early settlers' lives.[27]

In the Caribbean, a century of sailors dropping off pigs on any island they passed meant there were more pig settlements than human settlements. In 1609, when a British ship was wrecked on the then-uninhabited island of Bermuda, the crew was able to eat quite well thanks to the large herds of pigs. (In gratitude, once Bermuda was established, a boar appeared on its coins.) When the British began settling Barbados in 1627, while there were no humans, they found abundant feral pigs.[28]

While the proliferating pigs supplied explorers, settlers, and survivors of shipwrecks, they also offered provision for an increasing population of deserters, outcasts, privateers, and pirates of all nationalities. To cook the pigs, this rogue population adopted indigenous cooking methods. The Caribs were masters of smoke-drying meats and fish, a task they accomplished on a greenwood grate erected over a smoky fire. They called the grate a *boucan*. The French began to refer to those who adopted this cooking method as *boucanier*. When the English came along, they anglicized *boucanier* to *buccaneer*. Of course, this method of cooking recommended itself to other colonists, and its popularity spread. The Spanish, however, picked up the Arawakan word for the greenwood grill: *barbakoa*. It is from this that both the word and the tradition of barbecue would evolve.[29]

However, in the mid-1600s, things would change dramatically in the islands. Coffee, tea, and chocolate had reached Europe, which caused a dramatic spike in the demand for sugar. Owners of sugar plantations lost interest in raising pork—they could by this time buy salt pork from the mainland. Everything that could be cleared was cleared, including the forests and the herds of pigs the forests supported. Sugarcane took over the Caribbean.

The British controlled only the tiny island of Barbados in the early 1600s. When they took possession of the island, they populated it with indentured Scottish and Irish servants. Once freed from servitude, these disaffected Scottish and Irish workers would move to the Carolinas on the mainland, taking island ideas of cooking pigs with them and starting important pig-related traditions of their own.[30]

Though Britain would go on to take control of Jamaica (1655) as well, its primary involvement with pigs in the Caribbean would be as consumers. The bigger impact was the British-controlled settlements on mainland North America.

The British American Colonies

The English approach to the New World was different from that of the Spanish and Portuguese, though even between the main British settlements, there were differences. British colonists, on the whole, planned on being farmers, rather than soldiers and conquerors. The economy of Britain's North American colonies would be built on agriculture, and it was agriculture that would make these colonies powerful.[31] Farming had for so much of English history been the basis of not only the economy but of the culture that English settlers thought of farming as "the way things are." Civilized people had farms, raised crops, and kept animals. So that, of course, was what they expected to do in the New World.[32]

We who travel so swiftly and easily, and who live in a world where we expect rapid change, can hardly imagine how difficult adjusting to an alien landscape would be for people who required months to cross the ocean and spoke of change in terms of generations and even centuries. Aside from the need to produce enough food to survive, the desire to farm and raise livestock helped people make sense of their surroundings. In a world where everything was new, these were familiar. So, turning the New World's untamed wilderness into a land of charming farms reminiscent of the English countryside was the plan.[33]

Getting started was not easy. Two attempts were made to settle Roanoke Island, Virginia, the first ending in departure, the second ending in the disappearance of all the settlers, with their fates remaining a mystery to this day. The first successful British settlement was Jamestown, Virginia, founded in 1607. Life in the young colony was hard, and lives were generally short, but Jamestown did survive and began to grow.

Despite discouraging reports, hopeful settlers still came. There were compelling reasons that made facing the hardships seem worthwhile. Across Europe, unemployment and hunger were increasing. Even those who were not paupers had no opportunities for improving their lives. Escaping poverty, starvation, war, and persecution, along with the hope of material gain, were strong motivations for making a leap into the unknown.

Most students of American history will have read about how Jamestown and a second British colony, Plymouth in Massachusetts, settled in 1620, were saved by being introduced to corn (maize) by Native Americans, and that was certainly true. Corn was of immense importance, and would remain so as the colonies grew.[34] However, things didn't really begin to become less difficult until livestock began to arrive from England, as well as from other European countries and Spanish settlements to the south.

The survival of two colonies led to an even greater influx of those who were counting on the New World to give them opportunities the Old World could not offer. New settlements began to spread outward from the initial ones in Virginia and Massachusetts, and colonies began being discussed in terms of region: the Chesapeake and New England.

In the Chesapeake area, which included Maryland by 1634, as well as Virginia, adding pork and beef, though especially pork, to the diet of corn, contributed to reducing the threat of hunger. On top of this, the discovery of tobacco had given Jamestown an important cash crop, so trade with Britain increased—and more settlers came.

Life in the Chesapeake remained fairly rough throughout the 1600s. The death rate remained high, with a life expectancy among colonists of roughly forty years, a decade less than that of Native Americans in the area, who generally lived to around fifty.[35] Housing was rudimentary, and most livestock was expected to fend for itself in the surrounding forests—which created its own set of problems. When the British government took over the colony from the original investors, the Virginia Company, they claimed all "wild" livestock as property of the Crown. This created ill will among the colonists, who defended their ownership on the grounds that, since the animals had not existed in the New World prior to their arrival, they couldn't be wild. However, as livestock was increasingly allowed to run wild, ownership became increasingly difficult to prove, particularly for the rapidly growing numbers of swine, which often farrowed in the wild, with no witnesses as to which piglets were born to whose sows. Politics in the Chesapeake revolved largely around who owned what and when did an animal become wild, rather than semi-wild but still belonging to someone.[36]

While there were some commonalities between the colonies of the Chesapeake and those of New England, there were dramatic differences, as well. While Jamestown had been settled by men, Plymouth had been settled by families. That made a tremendous difference in focus. Jamestown was initially viewed as a get-rich-quick scheme, rather than a new home, and while families did eventually move to Virginia and Maryland, it took a while for that civilizing effect to have an impact. Also, the British government had given large tracts of land to those who paid their own way to Virginia, setting the stage for the plantation culture that would develop in the South, versus the villages and farms of New England.

In New England, while animals were allowed to forage, more livestock was kept on farms or grazed on the town commons of the tidy villages that had grown up. There was also a more concerted effort to improve the land. This primarily meant clearing brush and forest undergrowth, to make it easier for humans and livestock to move about and to protect domestic animals from the wild animals that filled the forests.[37]

Despite these and other differences, both regions came to depend heavily on pigs, as these were the animals that were most perfectly adapted to succeeding in a new environment. Forests had been dramatically reduced in Europe, but not here. Between the trees that grew wild and those intentionally planted by Native Americans, the newly settled areas offered a banquet for pigs. Oak and chestnut trees offered the type of mast pigs had lived on for centuries in the Old World, but added to this were North American trees widely planted by Native Americans, including pecans and hickory. So, as had happened in Europe centuries earlier, before widespread deforestation, pigs could be allowed to forage. This meant pigs were semi-feral, but it also meant they were easy and cheap to raise. This was more than just convenient; it was vital, because there was a tremendous labor shortage in the British colonies.

Aside from having abundant food, another reason pigs were popular is that the forests of North America were filled with wolves and bears, and other forms of livestock were not as good in combat as pigs are. Sheep in particular have no defenses, so they were easy targets for the big carnivores. With an easy, new source of food, wolf and bear populations grew. Cattle, if well fed and healthy, could generally survive because of their size, though they still did better on farms than in the wild. But pigs were (and are) formidable. Young pigs were certainly taken by wolves and bears, but mature hogs living in the wild have killed black bears.[38]

Menus revolved around pork. Bacon, ham, salt pork, and sausage became abundant. Fresh meat was enjoyed as well, and, besides a turkey, a nice roast of pork was a standard dish at early Thanksgiving feasts (because no one in the 1600s and 1700s would serve only one main dish at a big celebration, at least if it could be helped).[39]

However, abundance didn't mean ease. Pigs are destructive, so fields had to be fenced, or they'd be destroyed—not just food eaten, but the earth rooted up to a point where plowing became difficult. As farmers had done in Medieval Europe, so in the New World, farmers employed nose rings to keep pigs from rooting up fences to get to the fields. Official fence-watchers were appointed, and in time, the New England colonists created the office of *hogreeve*, a *reeve* being a person who enforces regulations (as in *shire-reeve*, which gave us the word *sheriff*). Hogreeves often had to assess damage, but their primary function was to prevent it.[40]

Having to deal with wandering hogs in New England, or rounding up semi-feral hogs in the forests of the Chesapeake, made pig raising a challenge for settlers. It also meant settlers were eating well most of the time. However, the introduction of livestock was shattering for Native Americans. It actually had a greater impact on their lives than the arrival of the colonists had. Pigs in particular would contribute to the destruction of the Native American way of life, but would also contribute to misunderstandings and conflict with settlers.

One of the biggest issues was food. All those wonderful nuts and acorns the pigs were enjoying had once been a source of food for Native Americans. But pigs didn't stop with mast. They ate everything. They consumed or despoiled beds of clams and oysters. They ate fruit and mushrooms. Deer and other animals the Native Americans hunted found the forests stripped of food. Indigenous wildlife was further hurt by the pigs' habit of eating young animals. As a result, the animals on which Native Americans relied started moving elsewhere. On top of that, Native American fields were not fenced, so pigs could also go after cultivated food. In 1666, one Native American pled with the Maryland legislature to let him and his people "know where to live & how to be secured for the future from the Hogs & Cattle."[41]

And yet the relationship between Native Americans and pigs was more complex than just hating them for eating everything. The colonists had originally thought that they could improve the lives of Native Americans by teaching them how to farm. Native Americans simply moved when their crops exhausted the soil, and they also followed migrating animals. If they had farms, the colonists reasoned, they could stay in one place. The colonists wanted them to adopt cattle, because that was very much the sort of animal that would make it necessary to put down roots. However, Native Americans had no concept of domesticated food animals. Cows were too alien. Pigs, they could understand, at least the feral pigs. They took almost no work to raise, and when one was needed, it was simply hunted. That was what Native Americans began to do. Unfortunately, while the colonists had often made gifts of pigs, the pigs Native Americans were hunting frequently belonged to colonists. Not understanding the concept of owning an animal that appeared to be running wild in the woods led to disputes. A tremendous amount of negotiating and explaining and asking for compensation for owners followed—because, on the whole, at least initially and especially in New England, the colonists actually wanted to remain friends with the Native Americans. Fortunately, Native Americans did have a concept of redress for damages and compensation for loss, so things remained civil for a while.

With the animals they'd traditionally hunted leaving, Native Americans became increasingly reliant on pigs for food. However, this could not last forever. More settlers arrived, and pigs continued to multiply. Some Native Americans moved farther west, but some thought they should at least express their anger. Cattle and sheep had become so identified with the colonists that Native Americans often killed and even tortured them, to demonstrate to the colonists their growing fear of the newcomers. They would not destroy the pigs, however, as they had become so dependent on them.

And still the colonists came. Most came because things were not improving in Europe, and people hoped they would be in the colonies. Not all came because they wanted to, however. The British shipped paupers and orphans to the colonies, along with political malcontents (which meant large numbers of people from Scotland and Ireland). As indentured servants from the Caribbean gained their freedom, many moved to the mainland. So, the population grew—but the amount of livestock grew faster.

In the mid-1600s, another major change occurred that led to more people moving to New England in particular. As previously noted, in the West Indies/Caribbean, all sources of local food were eliminated in favor of sugarcane. A fortune could be made by anyone who could supply food to the sugar plantations—and the burgeoning herds of pigs made that a reality for many. Salt pork became a hugely important export commodity for New England and, along with salt cod, became the "cash crop" that could parallel the success of tobacco in the Chesapeake. Everyone from Newfoundland to Virginia was already reliant on New England salt pork, but when demand spiked in the Caribbean, it became clear that farmers would no longer have to worry about simply subsisting.[42]

Of course, not all pig products were leaving the colonies. Pork and corn formed the basis of the diet for the entire colonial population. Ham or bacon were ubiquitous, no bean dish was made without salt pork, and the only cooking oil was lard. Corn made the colonists self-sufficient and pigs made them wealthy, but both kept them well fed.

By 1670, there were more than fifty thousand colonists in New England and more than forty thousand in the Chesapeake. The increased pressure on the land of so many people and still rapidly multiplying animals made it increasingly difficult for anyone to believe that colonists and Native Americans could live side by side. Trade became more complex. Native Americans became more reliant on European goods even as they were pushed away from European colonies.[43]

The Treaty of Westminster, signed in 1674, ended years of conflict between the British and the Dutch, and gave Britain the colonies that became New York and New Jersey. Then in 1681, King Charles II repaid a large debt he owed by giving William Penn a massive tract of land in the New World. Penn named the area for his father, and so was born Pennsylvania. Charles's brother, the duke of York (who would later become King James II, and for whom New York was named), then gave Penn the area that would become Delaware. This new region, between New England and the Chesapeake, would be known as the Mid-Atlantic.

Life remained difficult for settlers, not just because of conflict with the Dutch and quarrels with Native American, but also because the climate was brutal. The southern colonies were often plagued with severe droughts. The Little Ice Age continued to take its toll, with colonists and animals suffering and often dying during bitter winters. The effects of the winter of 1694–1695 underscore both the hardship and how great the herds had grown. In Maryland alone, during that winter, the cold killed twenty-five thousand head of cattle and sixty thousand swine.[44]

Thanks to hard work and salt pork, New England was getting wealthier and becoming increasingly independent, much to the annoyance of the British monarchy. What was worse, as far as the royals were concerned, was that New England traded with countries other than Britain. Britain began trying to exert greater control over New England trade, focus, and even religion. Britain's efforts were complex and unsuccessful.

New England did not remain alone in selling pork to sailors or shipping it overseas. The Chesapeake also got involved, producing barreled or pickled pork. Demand increased as supplies increased. By the late 1700s, tremendous quantities of barreled pork were being shipped out of both the Massachusetts and Virginia colonies, headed for Europe and the Caribbean. In 1774, Cuba, Madeira, and Portugal received sixty thousand barrels of pork from Virginia alone.[45] Of course, in 1774, the colonial period was about to end. The story of the pig would continue, but after the Revolution, it would be an American pig.

CHAPTER FIVE

American Pig

Shaping American Culture, Agriculture, and Foodways

The American Revolution gave colonists more than just independence from Britain. It gave them the possibility of living at something better than subsistence level. Land shortages had been a problem for most of the 1700s. In the northern colonies in particular, cities were crowded, with all the problems that entailed, especially the rapid spread of disease. However, there was nowhere to expand. By the early 1770s, the average New England farmer had only one-third as much land as the first generation of colonists, and nearly half of the population was no longer able to support themselves. People took what measures they could, from having smaller families to creating bartering systems to families supplementing their incomes by fishing, trapping, or working for other farmers. But it wasn't enough—because people kept coming, either voluntarily or involuntarily.[1]

After the French and Indian War (1754–1763)—a war in which colonists fought on Britain's side—colonists thought their problems were solved. The British had won from the French a huge swath of land west of the Appalachians, and colonists already had their eyes on valleys in Ohio. But then King George III issued a proclamation stating, among other things, that colonists could not move onto this land. The Royal Proclamation of 1763 also stipulated that anyone who had moved west of the Appalachians prior to the war had to move back east. (However, trading with Native Americans was not discouraged; it just needed a license now.)[2] This caused considerable anguish among those who had established farms in the Ohio River Valley, and it dashed the hopes of those who had planned to follow them.

The proclamation's intent was to reduce the frequency with which British soldiers had to come to the aid of British settlers who were attacked after moving onto Native American land. However, with cities crowded and not enough farmland to support the growing colonies, the colonists resented the proclamation, which appeared to them to simply be a new way of enforcing British control—an impression strengthened by ensuing taxes that punished them for purchasing food and goods from non-British sources. Parliament made it clear that the colonies existed for the benefit of the mother country, and Britain didn't want people becoming too self-sufficient. While the causes of the American Revolution were considerably more complex than simply bristling over the proclamation, it was a factor in the growing colonial discontentment—and it was an issue that was resolved by the Revolution.

The Growing Nation

Though burgeoning cities and shrinking farms were a bigger problem in New England than in the Chesapeake and Mid-Atlantic colonies, everyone had been running out of space. The American Revolution had given the colonists liberty, but, just as important for their future, it gave them the promise of land. With the land won from Britain, seemingly empty except for remaining French forts and trading posts, the new nation now stretched all the way to the Mississippi River. This region was dubbed the Northwest Territory, and one of the first orders of business was to start surveying it. The government needed to know what they were claiming, but they also wanted to ensure that the division of land west of the Appalachians was a bit more orderly than it was east of the mountains.

Thomas Jefferson had a vision of a nation of farmers, where people could have enough land to make themselves self-sufficient and independent. Plus, a lot of Revolutionary War soldiers had been promised land in lieu of payment for services rendered during the war, and the government wanted that to be fair and well organized. So only two years after the end of the American Revolution, the Ordinance of 1785 was passed, establishing a plan that would divide the new territory using a grid pattern, with land divided into townships that were six miles square and easily divided into 640-acre farms—a pattern that is still visible from the air when one flies over much of the Midwest. Surveying was already a respected job for educated and adventurous men. It was now vital, and surveyors would become key figures in the settling of the new country.

While some had jumped the gun and moved into the Ohio River Valley before the Revolution, settlers could now legally begin moving into the new territory. Then, in 1803, Thomas Jefferson accepted the French government's offer to sell its remaining North American claim, a large swath of land west of the Mississippi River. This transaction, known as the Louisiana Purchase, doubled the size of the United States. This became the New Northwest Territory. The nation now stretched to the Rocky Mountains. Florida, which fought on the side of Great Britain during the Revolution, had returned to Spanish ownership at the end of the war; the Spanish still owned much of the Southwest; and Russia, Great Britain, and Spain argued over the Oregon Territory, which included what is today Oregon, Washington, and most of British Columbia. It would be a while before "sea to shining sea" was a reality, but the infant government had plenty of land to deal with at this point. There were already plans for dividing the land, but how would one reach it?

Even before Lewis and Clark returned from exploring the continent, Thomas Jefferson had, with George Washington, begun planning a wide, modern road that would make it easier for pioneers to move at least part of the way west. Jefferson signed the legislation authorizing the construction in March of 1806. This would be the first federally funded highway in the new country. Called the National Road, it would in time become known as "The Road that Built the Nation." It was thirty-two feet wide—and it was paved. Construction began in 1811 in Cumberland, Maryland (earning it the additional name of Cumberland Road).[3] However, the project was soon interrupted by an event that would make settling the new territories seem vital. In 1812, the British government, which felt that it was still the rightful owner of the colonies and that everyone in the United States was a traitor to Britain, began arming disaffected and displaced Native Americans in Canada. These indigenous recruits were to aid British troops in punishing the renegade colonials and retaking their property (though Britain intended to retake it for themselves, not for the Native Americans). These forces swept down through the Great Lakes region and headed east, where the British Navy would attack from the sea. Settlements were wiped out. Forts were captured. Washington, D.C., was burned to the ground, but the redcoats were finally stopped in Baltimore. The British had failed to reclaim their colonies. The War of 1812, sometimes called the Second War of Independence, created a sense of urgency. People felt that safety lay in settling the region as quickly as possible. Once the war ended, work resumed in earnest on the National Road. Over the next two and a half decades, the road opened up the west, stretching into the Northwest Territory, terminating in Vandalia, which was, at that time, the state capital of Illinois.[4]

The National Road was far from the only way folks were moving west, and not the earliest way. The first people to head into the wilderness were the backwoodsmen, who hiked into the unknown with nothing more than what they could carry, which usually included at the very least a hatchet and seed corn for planting. For those who preferred traveling by water, the Ohio River offered relatively easy access to the area that would become the states of Ohio and Kentucky. The Great Lakes offered another route into the region, as did tracks blazed by earlier explorers.

With surprising speed, people spread across the new land, driven by hope for a better life and the promise, unthinkable in the Old World, of owning land, of getting ahead by hard work and not because of some accident of birth. The new land meant danger and hardship, but it also meant possibilities, the most urgent possibility being not starving to death, but also the possibility of someday doing better than anyone could have imagined in the lands they'd left behind. And a lot of that doing better revolved around eating.

Developing Foodways in the Growing Country

Food on the frontier could be rough, but it wasn't around for long, as the frontier disappeared so rapidly. Within fifty years of the end of the War of 1812, westbound Americans had settled all the way to the Rockies and the Transcontinental Railroad was well under way. However, the period did witness the early development of American foodways.

It would hardly be an exaggeration to say that corn and pigs built the country. These two helped the settlers of the eastern states survive and thrive in the earliest years, and now they did the same on the frontier. No one wanted to stop with corn and pork, but it was definitely where things started.

Americans already had nearly two hundred years of food traditions in place by the time westward expansion became possible. Maize, or Indian corn as it was still known then,[5] had been wholeheartedly adopted, but corn was good for more than dining; it was ideal for feeding animals. In fact, feeding corn to livestock had become the American agricultural paradigm. Corn was easy to grow and produced abundantly, and, as a result, meat became abundant—and was abundantly consumed.

Average annual meat consumption by 1800 ranged from 150 to two hundred pounds per person.[6] Even servants were fed meat. Meat supplied the calories needed for hard labor and cold weather. (The Little Ice Age wouldn't finish its retreat until the mid-1800s, but even then, the northern part of the country still had to contend with long winters.) Eating meat was also a way of preventing pellagra, the devastating niacin-deficiency disease that haunted regions that depended entirely on corn. And the easiest, most prolific source of meat was pigs. Beef was more highly valued, but it was meat for the wealthy or for holidays. Pork was everyday food.

Lewis and Clark had among their stores at the beginning of their journey of exploration thirty-seven hundred pounds of cured, barreled pork and seven hundred pounds of lard, as well as six thousand pounds of corn.[7] While the earliest backwoodsmen may have relied on their hunting skills, taking only corn to plant, most of them soon had pigs as well. Archaeologists excavating Davy Crockett's Tennessee birthplace found that more than 92 percent of the bones unearthed there were from pigs. Even Abraham Lincoln, the son and grandson of backwoodsmen, described himself as a "mast-fed lawyer," comparing his upbringing to that of the semi-wild, foraging pigs that would have provided his family food.[8] When covered wagons began heading west, they generally carried seventy-five pounds of bacon per adult.[9]

But pioneers didn't just pack bacon. They took pigs. Pigs were the natural choice when moving into new areas, because, like corn, they could thrive just about anywhere. The pigs that headed west were long-legged, short-bodied, rough-coated creatures that could take care of themselves. Their feistiness earned them such nicknames as "stump-rooters," "snake-eaters," and "wound-makers." Piglets might be carried in the wagons, but the tough, adaptable adults could just trot alongside their owners—or head off to forage. Remarkably, with the whole expanse of wilderness before them, the pigs came when called.[10]

Foraging pigs continued to eat whatever they found—tree nuts, fruit, roots, grasses, and small animals—and, as had happened back east, so on the frontier, Native Americans often lost their food sources because of that foraging. The thing pigs loved best was corn—which was a disaster if they found a cache of Indian corn.

Once farms were established, pigs' love of corn was a blessing for farmers, as feeding corn to pigs was one of the easiest ways of getting corn to market. It was this that would in time earn pigs the new nicknames of "cornfields on legs" or "corn on the hoof."[11]

Corn and hogs, also sometimes rendered "hogs and hominy," became a common description of the farming paradigm that was spreading westward. So iconic was this pairing that in the 1830s Tennessee became known (temporarily) as The Hog and Hominy State, as its production of corn and pork was so tremendous.[12]

For farm families looking to settle, pigs were for the family to eat, or possibly trade. A cow might be brought along for milk, but beef was not on the menu. Pork and corn would appear at three meals a day (or more).

Even recipes that didn't appear to be pork recipes usually included salt pork as a seasoning.

Their ready adjustment to new surroundings was an advantage, but pigs were also ideal for families on the frontier because pigs are reasonably easy to process with only a few people available for the work. Once a person had neighbors, work (and meat) could be shared, but initially, families had to be self-sufficient.[13]

Culinary refinement would trail behind settlement, at least until survival was assured. After visiting the United States in the 1830s, British novelist Frederick Marryat returned home to report on the state of food in the young nation. He wrote,

> The cookery in the United States is exactly what it is and must be everywhere else—in a ratio with the degree of refinement of the population. In the principal cities, you will meet with as good cookery in private houses as you will in London, or even in Paris. . . . Of course, as you advance into the country, and population recedes, you run through all the scale of cookery until you come to the "*corn bread, and common doings*," (i.e. bread made of Indian meal, and fat pork) in the far West. In a new country, pork is more easily raised than any other meat, and the Americans eat a great deal of pork.[14]

American settlers were from several European countries—and increasing numbers of countries as the years went by—but foodways initially were more linked to necessity than culinary tradition. Fortunately, while corn was new to many, pork was a major part of most people's traditions.

There really weren't clearly defined national cuisines at the time that the United States was getting started. Until the "middling class" began to emerge in the 1800s, food was not defined by nation so much as by class.[15] As a result, what was available shaped American foodways more than inherited ideas of what food should be, other than general ideas of meat and starch. Later European immigrants would introduce the clearly defined cuisines we now associate with various nationalities—cuisines that in many cases had been shaped by foods introduced from the New World. But in the early 1800s, while there were fish and game, American cuisine was developing based largely on the two most reliable items: corn and pork. Once people were settled, gardens would offer apples, onions, peas, and other fruits and vegetables brought from Europe, but pork and corn would still be the foundation. Corn would become so important that an entire region came to bear its name: The Corn Belt. And pork, the most important meat in the country since the 1600s, would remain the primary protein source until the early 1900s.[16]

Which is not to say people didn't bring ideas from Europe, especially Great Britain. However, much European cuisine, especially as relates to pork, was still guided by ideas inherited from the Celts (bacon and ham) and the ancient Romans (roasts and sausages). These traditions were carried to the New World and took hold here, as well. Salt pork would remain vital into the twentieth century, both as a means of survival[17] and as a seasoning ingredient in too many recipes to name. Lard was the cooking medium for nearly everything. So many came to the United States to escape poverty, one might say frugality was imported from Europe, and there was a great passion for what we now refer to as nose-to-tail cooking. Pigs' trotters were delicacies. Pigs' heads were boiled, stripped, and molded into souse, brawn, or headcheese, the name depending on the preparer's background but the dishes virtually identical. And a recipe from the late 1800s states that, after roasting a young pig, one should "chop the liver, brains, and heart small," and add them to the gravy.[18]

Every part of Europe introduced sausages, though German-speaking countries introduced a greater variety than any other group. Cleaned intestines of various farm animals made up what we now call "natural casings." The large intestines always went to sausage making. The small intestines were used for sausage as well, but part would be set aside to make chitterlings.

The word *chitterlings* first appeared in Middle English in the 1200s, though the consumption of innards was fairly universal in Europe. Even today, chitterlings are popular in France, where they are stuffed into the sausages called *andouillette*. The Southern chitterling tradition, however, came to North America from England. Chitterlings were being enjoyed in Virginia by 1627. By the time the plantation system was well established, hog butchering was a major annual event. Whites and blacks shared the work and shared the chitterlings—though they dined separately. The contracted form of chitterlings, *chitlins*, did not come into use until 1842, but it is now commonly seen in American texts, especially those describing Southern cooking. It was only in the second half of the twentieth century that dining on chitterlings gained its current associations with African American culture.[19] All that said, enslaved Africans played a big part in the development of cuisine in the South, primarily because they did so much of the cooking.

One unintentional impact of the slave trade was that peanuts arrived in the South with African captives. Peanuts were indigenous to South America, but they had been taken to Africa by Spanish and Portuguese explorers who thought they would be good, cheap food for the Africans they would employ and enslave. Then, in the 1700s, when Africans were being transported to

Virginia, they brought peanuts with them. Initially, peanuts were considered appropriate only for livestock and the poor.[20] While people's opinions of peanuts would improve, feeding them to pigs led to the creation of the South's best hams, including the almost legendary Smithfield country hams.

It is worth remembering that both North and South were settled by people from essentially the same parts of Europe, most especially Great Britain. British settlers brought with them their culture, language, and ideas of government and law (most of which the United States adopted), but they also brought their ideas of food. They introduced (as noted above) chitterlings, but also brought their love of greens, including collard greens, which they planted widely, plus all the roasts, stews, pies, custards, sausages, and soups that a few thousand years of trade, migration, and invasions (from ancient Romans to the Norman Conquest) had developed. Even as people increasingly identified themselves and their foods in relation to a state or region, the first American cookbooks were essentially British cookbooks with a few new foods added.[21]

Even with input from enslaved Africans, who introduced, most notably, okra, sesame, and black-eyed peas, the food culture in the South didn't differ dramatically from that of the North, at least initially. Everyone ate pork and corn, even if it was in some cases supplemented by other foods. As in Europe, with few exceptions, money and location had a bigger impact on what one ate than any regional identity. In New England, seafood would supplement pork and corn (especially when there was more money to be made shipping salt pork to the Caribbean), and the wealthy added such luxuries as beef and tea to the menu.

That began to change after the Civil War. The North, which had won in part because it had more food than the South, became more industrialized and wealthier, while the devastated South struggled simply to survive. The North was also more urban, as opposed to the South, which was largely rural. The foods that became associated with the South were the foods that the poor everywhere had eaten previously, but now Northerners could leave behind the tougher greens and rougher cuts of pork, and could even begin to buy more beef. Of course, in the North, as people flooded into the cities, they could no longer rely on what they could grow themselves, while large, varied vegetable gardens remained a major source of food in the South. Often, rural Southerners, including African Americans, though poor, had access to a greater variety of fruits and vegetables than people in the urban North.[22]

Even as people moved into the western lands, there would be little change to diets, simply the adoption of new foods encountered along the way, either from collisions with Spanish settlements or the influx of people from different European countries. The Cajun and Creole cuisines of Louisiana were an exception, as they developed outside the British influence of the eastern states. However, it wasn't until the twentieth century that people began seriously trying to attach certain foods to certain regions or groups.[23] It was all just American food.

When speaking of "what was available," it is worth noting that this involved more than just the meat and vegetables one could raise oneself. The first cast-iron kitchen stove small enough for a private home wasn't invented until 1834. People simply cooked in their large, open fireplaces. That meant preparation methods were fairly limited: spit roasting, a big pot of stew, and something baking in a Dutch oven nestled in the hot coals. It wasn't just limiting, it was dangerous, as women wore long skirts and catching fire was more common than one would like to think.

Canning wasn't available until 1809—and that's canning in glass jars. French Chef Nicolas Appert was awarded twelve thousand francs for inventing this preservation method, which was considered vital to the success of Napoleon's army. Everything imaginable was canned: meat, seafood, fruits, vegetables, even soups. Canning soon jumped from France to England, where the focus was on lining iron cans with tin, rather than using glass (though still with a goal of feeding the military). Tin cans weren't fully developed until 1820, and even then, they were cut and welded by hand. From England, canning, both in glass and tin, jumped to the United States. Initially, with welded tin, the biggest problem was opening the cans, as the first can opener wasn't invented until 1858.[24]

By the beginning of the Civil War, canners (still welding by hand) were able to supply food to the Union Army. Soldiers returned home with enthusiastic tales of food that would never spoil, and suddenly everyone wanted food in cans. Canners flourished. The oldest trademark in the United States that is still in use dates to this period—and involves pigs and cans: Underwood Deviled Ham, still available in grocery stores today, was created in 1868 and trademarked in 1870.[25]

The popularity of canned foods was so great and growth was so explosive that some people began to worry about quality and safety. But even with concerns about quality, canning made so many more foods available, and available throughout the year (as even on successful farms, food would always be running low by the end of winter) that no one was going to be dissuaded from accepting canned foods. Add the switch from the limitations of cooking in an open fireplace to having a stove with multiple burners and an oven, and cooking was transformed. Some missed spit roasting, but most celebrated the greater number of cooking options—and delighted in serving summer vegetables in the middle of winter.

Interestingly, another impact of the Civil War was an explosion in the production of cookbooks. In order to raise money to buy supplies, clothes, and bandages for the troops, women began collecting recipes from friends and neighbors and creating cookbooks to sell. So, between roughly 1820 and 1865, the world of home cooking was not only transformed but also more widely communicated.[26]

Changing Attitudes, Changing Options

Cities were growing quickly in the United States. New York grew by nearly 750 percent between 1810 and 1860—faster growth than the still-small municipal government could handle. Garbage and pollution were becoming serious problems. And New York was still entirely on Manhattan Island at this point.[27] In addition to the human population, by the 1820s, there were roughly twenty thousand hogs in Manhattan. Relied on by the urban poor, these hogs were raised on the streets. As had been the case in Europe, the fact that pigs would eat garbage made raising them almost free. However, having thousands of large, unruly animals foraging in the gutters and alleys of the city was hardly wholesome and was often dangerous. Local authorities cited not only health issues but also issues of safety, especially for small children. New York was not the only city with pigs, but as the largest city in the country, New York was the first to be forced to deal with it.[28]

Processing the pigs further compounded the problems. In 1850, local butchers in New York were slaughtering close to ten thousand animals per week—cattle, sheep, and pigs—to feed the growing urban population. Because many New Yorkers were becoming increasingly well off, interest in "snout-to-tail dining" was waning. Organ meat was simply dumped by the ton into rivers, though some people found ways to make money processing the offal as well as other unwanted materials. The area between Sixth and Seventh Avenues became known as Hogtown, an area devoted both to piggeries, where pigs were raised and fattened, and to the processing of bones, skin, fat, and offal. Unfortunately, as Frederick Law Olmstead noted when he toured Central Park in 1856, the park's grounds "were steeped in [the] overflow and mash of pig sties, slaughterhouses, and bone boiling works and the stench was sickening."[29]

Dealing with the urban pig population led to conflicts that ranged from riots to what bordered on combat, because, despite the obvious disadvantages, the poor were loath to give up their pigs. The Piggery War of 1859 was the first major effort to eliminate the Manhattan pig industry, though the last of the urban pigs would not leave Manhattan until after 1910. However, things did begin to improve.

The problem of feeding the eastern cities was a big one. In the late 1860s, when the population of Manhattan was nearing a million, it took 1.1 million animals each year to provide the needed meat. Even as more animals were raised outside the city, they were still brought to the city for slaughter and processing. The opening of the Midwest was part of the answer to the problem.[30]

Not all who were wandering westward were in search of a little house and enough to eat. Some of those establishing themselves on the frontier were looking for a lot of land, in order to raise enough pigs to feed the growing cities back east. Settlers with really large herds had to keep moving west, because foraging pigs needed space, with the ideal population density being two people or fewer per square mile.

Of course, pigs being raised to feed the cities meant two-way traffic. In fact, hogs were moved in stunning numbers along the tracks and turnpikes—five times as many as all other animals combined. In 1847, one turnpike tollgate in North Carolina recorded the passage of 692 sheep, 898 head of cattle, 1,317 horses, and 51,753 hogs. The big hog drives moved the animals along trails that would become well worn, from Ohio to Maryland, Kentucky to Virginia, Tennessee to Alabama.[31]

So, pigs crowded the roadways and spilled into the cities, to feed the urban multitudes.

By the 1830s, Americans were beginning to build railroads, so animals had a way of reaching the city by means other than walking. However, they still had to be moved to the slaughterhouses, so there were still vast numbers of animals moving through the streets—and huge amounts of blood and guts that had to be disposed of. Americans, while still flocking to cities, and still demanding abundant meat, were ready to give up having their meat processed in their urban neighborhoods.

It was the perfect time for something new, and new was what the Heartland would offer—not just a new place, but new ideas, new ways of handling things. Moving livestock via railroads was the beginning of change. Then, after pork packer George Hammond introduced the refrigerated train car in 1868, meat, rather than live animals, could be shipped.[32]

New York butchers and meat packers fought the changes at first, as did those who had gotten wealthy off the shipping of live animals and processing of meat. When the Swift brothers began shipping fresh meat from Chicago, the Vanderbilt family of New York promised to destroy them. But people soon realized that not having great herds of livestock walking through the streets and animals being butchered in every neighborhood actually meant a cleaner city. The Swift brothers won.[33]

The isolation of the frontier vanished. By the end of the 1800s, railroads and the companies that ran them had grown to almost unimaginable importance. There were, by then, more than two hundred thousand miles of tracks sprawling across the nation, connecting rural towns, big cities, seaports, and stockyards in a giant web of enterprise. Anything could be, and was, shipped, including vast amounts of livestock.[34] The country would increasingly look to the Heartland for its food.

CHAPTER SIX

Corn Belt/Hog Belt

How Hogs and Hominy Helped Define a Region

In an 1862 speech, Abraham Lincoln roughly defined the parameters of the region that would come to be known as the Middle West: "The great interior region, bounded east by the Alleghenies, north by the British dominions, west by the Rocky Mountains, and south by the line along which the culture of corn and cotton meets."[1] Like Thomas Jefferson before him, Lincoln envisioned it as a region of productive farms that would feed the country.

The settlement of the region began nearly a century before the terms Corn Belt and Middle West came into existence, but the identity of the region was well formed by the time the names were coined in the late 1800s.[2] The Corn Belt and Middle West (now more commonly called the Midwest) are not truly identical. The "belt" runs over Ohio's border into Kentucky and begins to thin and vanish along the northern and western edges of a few Midwestern states. However, the two regions are so nearly the same, not only in location but also in history, culture, and economy, that one cannot discuss one without discussing the other.

One of the things that defined the Corn Belt was feeding corn to livestock. As a result, a large part of the region picked up the additional nickname of Hog Belt, because regardless of whatever else was being raised, wherever there was corn, there were pigs. Of the top ten pig-raising states in the United States, only two are not in the Midwest: North Carolina and Oklahoma—and Oklahoma borders on the Midwest. Iowa is, and probably always will be, the top pig-raising state, just as it is the top corn-growing state. In fact, Iowa's pig production is greater than the combined production of the second and third top states.[3]

But how did this come about?

Creating the Heartland, 1783 to 1861

The Renaissance concept of *res nullius* may have made it easier psychologically for settlers to move into the newly opened lands, but it didn't make it easier physically. There were no roads initially, the terrain was rugged and uneven, and the woods were filled with bears, wildcats, and wolves, as well as unhappy Native Americans who were well armed after the War of 1812. But the land was fertile and, for the hopeful settlers, there really were no good alternatives to moving westward. Europe offered poverty, war, persecution, and famine, and the eastern states, poverty and overcrowding. The land to the west looked empty to Europeans. At this point, official government policy was still to respect the legal and political rights of Native Americans, so when settlers saw land that clearly belonged to Native Americans, they offered money and trade goods, believing that the concept of buying land would be universally understood. Efforts were often made to live together peacefully, but with settlers increasing in numbers, and with pigs regularly locating and eating Native American caches of corn, Native Americans saw their traditional lives vanishing. Some resisted, but the indigenous people of the Midwest were either bought out or defeated and moved out, pushed farther west.[4] The frontier offered hope to new settlers, and that hope led to an unstoppable tsunami of Europe's "huddled masses yearning to breathe free."

Philosophers, scholars, and writers of the early 1800s idealized the frontier, viewing it as a return to Eden. According to Norman Foerster, Ralph Waldo Emerson once said, "Europe stretches to the Alleghenies; America lies beyond."[5] While this idea overlooks the importance of the eastern states, it does recognize that something new was beginning with westward migration. In a very real sense, the Northwest Territory, as it was then called, was where the images would develop that are most strongly associated with what it means to be American. And at a time when almost everyone had to farm in order to survive, the pursuit of happiness meant owning land.

Optimism was probably further encouraged by the fact that the Little Ice Age was winding down, finally ending in the mid-1800s. Shorter, less brutal winters would make life—and agriculture—easier. It seemed as though nature was in agreement with the philosophers of the age.

Right from the start, people recognized the potential of the land itself, not just the ideals or desperation that propelled settlers westward. The new territories were broad and beautiful, with lush valleys, numerous rivers, and abundant forests. Even the realists admitted that this was the perfect place for farming—and a phenomenal place to raise pigs. In 1864, veterinary surgeon Robert Jennings published a practical work on how to raise and care for livestock. Early in the chapter on "American Swine," he waxes rhapsodic about the region:

> On the rich bottoms and other lands of the West, however, where Indian corn is raised in profusion and at small expense, [swine] can be reared in the greatest numbers and yield the largest profit. The Scioto, Miami, Wabash, Illinois, and other valleys, and extensive tracts in Kentucky, Tennessee, Missouri, and some adjoining States, have for many years taken the lead in the production of Swine; and it is probable that the climate and soil, which are peculiarly suited to their rapid growth, as well as that of their appropriate food, will enable them to hold their position as the leading pork-producers of the North American Continent.[6]

Jennings's work recognizes another key to Midwestern success: corn. Corn-fed pigs tasted better than pigs that had been raised amid urban alleys and gutters. As a result, corn-fed pigs brought a price that was 30 to 60 percent higher than foraging pigs.[7] As noted previously, the flavor of the things pigs eat affects the flavor of the pig.[8] Animals raised in forests are gamier tasting than those raised on farms. Those raised on corn have sweeter flesh. The fat of corn-fed pigs is also sweeter—and at a time when everything was cooked in pork fat, that was important.[9] Corn also made the flesh firmer, and so meat of corn-fed pigs was deemed superior to that of mast-fed pigs.[10]

Though people had been wandering into the Ohio Valley since before the American Revolution, the true initial period of settlement of the Midwest was bracketed by independence and the Civil War. Ohio became a state in 1803, and by 1861, it had been followed into statehood by Indiana, Illinois, Missouri, Michigan, Iowa, Wisconsin, Minnesota, and Kansas. And that was pretty much the trajectory, east to west, of the beginning of the Corn Belt.

It was in Ohio, in the Scioto and Miami River Valleys, that the Corn Belt is said to have been born. Ohio is also where serious efforts began to breed pigs suited to the region.[11]

In the early 1800s, the British were busy breeding pigs for their own needs, and Americans began to import some of the new English breeds. However, Americans knew they couldn't just stop with what the British had done. With the seemingly endless land spreading out before them, Americans wanted the largest pigs possible, while the British were going for breeds that were somewhat more compact. Americans also wanted lots of fat on their pigs.

Sometime between 1840 and 1846, farmers in Ohio's Miami Valley (though precisely who is uncertain) bred a pig that would become the success story of the era: the Poland China. (English pigs known as Big Chinas had been introduced into the area in 1816, to improve the local stock. Local pigs with Big China genes were crossed with imported Berkshires and Irish Graziers, creating the Poland China. Because the breed's pedigree wasn't registered until 1876, the origin of the "Poland" part of the name got lost, though some suggest the original breeder might have been Polish.) The Poland China was big and meaty, with abundant fat, but equally important for the time was that, even at market weight, it was still able to walk the often-considerable distance to market.

The Poland China was the first American breed, and it became very popular in the Corn Belt, because it fattened well on corn. However, it was not the only American breed developed. Pigs were also imported from Asia, South America, and across Europe to add to the gene pool. Back east, the Chester White was developed along the Delaware River and the Duroc was bred in New York.[12]

Developing hand in hand with the growing adoption of the corn-livestock paradigm and breeding of pigs was a new concept in livestock handling: the feedlot. Cattle had joined pigs as the Scioto Valley became more settled, and in roughly 1800, the new concept was taking shape: don't feed livestock in a barn, feed them in an open lot. Husked corn, silage, and other fodder would be fed to the cattle in an open lot of eight or ten acres. Then, because much of the food would either be untouched by the cattle or pass through the cows undigested, the cattle would be moved to another lot, and the hogs would come in and clean up everything left behind, including the cellulose-rich cow dung.[13]

Settlers had continued to spread westward even before the Poland China was developed, but the new breed soon followed their trails. The Poland China was designated a lard-type pig, rather than the meat-type pig more common today. It gained weight readily from corn, and, in a day when there were few transportation options, it became the region's most popular way to get corn to market.[14] In addition, the market demand worldwide for lard was huge, and this new breed produced it. This and other lard-type breeds made the Corn Belt the center of global lard production. Lard was not only the most important (and sometimes only) cooking fat, it was also the top industrial lubricant, since petroleum had not yet been discovered.[15] And this was worldwide, not just on the American frontier. Also, fat made meat easier to preserve (fat on corn-fed pigs, that is—fat on acorn-fed pigs goes rancid more quickly). In addition, most of the pork at that time was being brined, barreled, and sold to feed miners, sailors, and slaves, who most likely needed as many calories as they could get to stay fit for work.[16]

And so settlers continued westward, across Illinois and Iowa, converting grasslands to cornfields and cornfields into pork and lard. Towns and cities soon joined the farmers on the rapidly developing prairies—though some settlements predated this period. More than a dozen towns in the Midwest had their genesis as forts or trading posts created by the French in the 1700s. Fort Detroit was created in 1701.[17] St. Louis was a fur trading post founded in 1764.[18] However, most towns in the region would grow after the American Revolution, some becoming cities that would play important roles in the pig story.

Cincinnati was founded as a village in 1788, fifteen years before Ohio was made a state. By the 1820s, it had become the biggest city in the Corn Belt. Though the town's beginning was rocky, thousands of settlers flooded in, basing their hopes for success on the considerable traffic along the Ohio River, which, because it flowed into the Mississippi River, gave the Midwest access to the bustling port of New Orleans. There was always work in places like this—seasonal work, but long seasons. The farming communities and herds of pigs had grown so rapidly, and Cincinnati was so ideally placed for river transport, that the city became the destination of farmers from a quickly widening area. It was in Ohio, at this time, that the concept of fattening the pigs after they reached their destination was developed. At first, the herds came from only a few miles away, but by the 1840s, pigs were being driven more than a hundred miles to Cincinnati.[19]

Pig farming was readily adopted by newcomers because it took less expertise than other types of farming and gave the highest return for work done. Pigs cost little to raise and produced meat quickly. They were ideal for small farmers, as well as for more ambitious ones.[20] And so, as Ohio settlement grew, the numbers of herds grew.

In Cincinnati, having pigs arriving in large numbers created a demand for corn, to fatten the pigs, and packinghouses (so called because most pork was still packed into barrels), to process them. Salt was also needed, and it arrived by the ton from saltworks back east.[21] The first recorded packinghouse in Cincinnati started business in 1818, though people had been slaughtering and packing pigs on a more modest scale before this. By 1820, Cincinnati was packing 114,000 barrels of pork a year.[22] The system became more sophisticated—and faster—as time went by. Butchering a hog was a daylong process for a family, but on one of Cincinnati's "disassembly lines," it took less than a minute. As landscape architect Frederick Law Olmsted noted when he visited a Cincinnati packinghouse, "Amazed beyond all expectation at the celerity, we took out our watches and counted thirty-five seconds, from the moment one hog touched the table until the next occupied its place."[23]

The number of hogs processed continued to grow. Cincinnati became the country's leading pork processor, earning itself the nickname "Porkopolis." Before long, pigs were coming to Cincinnati from Indiana and Kentucky. And still, Cincinnati struggled to keep up with demand. Cities kept growing, and cities wanted meat. The United States focused on transportation systems, not to move people, but to move food. Three thousand miles of canals were created in the 1820s and 1830s. Seven thousand miles of railroad tracks were laid in the 1830s and 1840s. By the 1850s, packers in Cincinnati were processing an average of 334,000 hogs per year. And still, meat shops were sometimes empty, because the population was growing at such a pace that demand was greater than the capacity to produce meat.[24]

Meat and lard were not the only things coming out of Cincinnati. In 1837, two men, William Procter from England and James Gamble from Ireland, shook hands and became business partners. And so began the Procter & Gamble Company, purveyors of fine soaps and candles—products that relied on tallow and lard from the nearby pork processing operations. Procter and Gamble researched ways to make better candles and finer soaps, but in the 1840s, the company's best-selling product was lard oil. Far inland from the oceans, lard oil was easier to come by than whale oil, and so lard oil—oil made from animals now sometimes called "prairie whales"—was fast becoming the top oil for burning in lamps. Procter & Gamble would grow to be the world's largest consumer goods company, but its start was anchored in the pigs of Porkopolis.[25]

And Cincinnati, though it was the largest center of pork processing in its day, was not the only city with pork-focused businesses. St. Louis, too, processed pigs—and drew people to the city who would take advantage of the byproducts of pork packing. In 1843, a German soap manufacturer named Eberhard Anheuser left Europe and headed for St. Louis. While he would later join up with another German, Adolphus Busch, and get involved in making beer, his first fortune was made creating the largest soap and candle company in St. Louis.[26]

Chicago was founded in 1833, and much of the surge in transportation creation focused on this city by the lake. The opening of the Erie Canal in 1825 had made it possible for the Midwest to ship goods easily to the East Coast (and vice versa). The Illinois and Michigan Canal, completed in 1848, gave Chicago access to the Mississippi River and New Orleans. As railway lines were laid, a lot of them terminated in Chicago, where they had access to the shipping facilities on Lake Michigan and the Chicago

River, because moving goods by water was still more cost-effective than any other method, including trains.

With trains available, the big hog drives began to taper off. Breeding programs began to focus on flavor, rather than on endurance, because pigs no longer needed to walk to market, just to the train station. There were still issues in all cities with huge numbers of pigs making the trek from train to slaughterhouse. The noise and traffic involved in moving large herds was something with which city dwellers began to lose patience. Then there was the issue of the filth from the pigs, not to mention the inconvenience of having gardens and lawns rooted up by strays. Chicago would succeed in solving this problem, but not until after the Civil War.

Still, hogs kept increasing in number. By 1860, hogs in Illinois numbered more than 2.5 million, and in Iowa, were approaching a million.[27] But with the beginning of the Civil War, numbers would grow even faster.

The Civil War and Lincoln's Legacy

While Cincinnati was technically in the North, there were many in the city who sympathized with the Confederacy. Plus, both secession and conflict would disrupt business and transportation along the Ohio and Mississippi Rivers. Processing would continue in Cincinnati, but much of the pork industry, plus millions of pigs, now got detoured to Chicago, where trains kept meat moving, even during the war. Chicago was about to become the new Porkopolis.[28]

The only thing greedier for meat than a growing city is a moving army. During the Civil War, pig production in the Midwest exploded, because barreled pork was the easiest way to feed the troops. Pig drives grew in size, as animals were moved to wherever the front was. Most of the Midwest was in the North. (The only exception was Missouri, which was one of the "Border States," states that sent volunteers and supplies to both sides of the conflict, and actually fought internally, but didn't secede from the Union.[29]) Because the Midwest had the majority of the nation's pigs by this point, the Union Army had plenty of meat.

However, meat wasn't the only thing the Union Army needed. It also needed volunteers. The region had grown at an unprecedented rate, and by the start of the Civil War, there were more than seven million people living in the "Old Northwest." These states would provide remarkable numbers of volunteers in response to native son Abraham Lincoln's call. Minnesota became the first state to form a volunteer regiment in response to Lincoln's request. Only New York and Pennsylvania sent more volunteers than Ohio,

Illinois, and Indiana, and Missouri, Iowa, Michigan, and Wisconsin all ranked ahead of most eastern states. Percentages are even more impressive, with 57 percent of Indiana's military-aged men going to war. Illinois came in second. With both food and men to offer, the Midwest would have a tremendous impact on the outcome of the war.[30]

Of course, sending most of the military-aged men to war meant sending most of the farmers to war. Part of the labor shortage this created was made up by women and children. Plus, while immigration slowed somewhat during the conflict, it didn't stop, and 430,000 new farms were established in six Midwestern states even as the Civil War continued. In addition, all the transportation systems, the gains in processing speed, and efforts to modernize farming that had occurred in the Midwest meant that it was actually possible to ramp up production even with fewer farmers. Much of this had been put in place to feed the growing cities, as the number of farms had steadily declined, but now it made survival in wartime possible.[31]

While the North generally had a steady food supply, the South was not so fortunate. The big hog drives that preceded the war (and would commence again afterwards) provided much of the meat eaten in the South, and this source was dramatically reduced.[32]

A hog shortage was not the only problem the South faced. Shipping blockades kept salt from reaching the Confederate states, and without salt to preserve the meat, the hogs that were available were only useful for a few days after slaughter. By 1862, the salt famine was hurting everyone, and a great race was on to try to find local salt. People near the coast cooked food in seawater. Mines and salt-processing operations were started and were as often targeted by Union forces, who knew precisely how important salt was. On Avery Island, Daniel Avery and Edmund McIlhenny (who would, during Reconstruction after the war, create Tabasco sauce using chilies grown on Avery Island), discovered large salt deposits, and so, in 1863, the Union Army took control of Avery Island. No salt meant there was rarely meat, and hunger began to be an increasingly serious problem for not only the Confederate Army, but for everyone in the South.[33]

At the same time the South was starving, the Midwest was exporting tons of food to Europe, which helped replace any financial losses due to not having the South as a customer for wheat and pork. While food alone cannot win a war, soldiers weakened by hunger, or absent because they'd left to care for their starving families, reduced the effectiveness of the Confederate forces.[34]

By the end of the war—and largely why the war ended—Grant and Sherman had decided that increasing Southern hunger might be the best way to bring the conflict to a swift conclusion. Sherman, who coined the famous phrase, "War is hell," understood that the South would fight as long as they could. He was determined to make sure they couldn't. Union troops were instructed to live off the land and destroy everything they couldn't eat themselves. During his "March to the Sea," Sherman's troops destroyed millions of tons of corn and fodder and untold numbers of hogs and other livestock. By the time Sherman was through, the South was in ruins—and ready to surrender.[35]

Reuniting the nation and rebuilding the South would be a long, slow, painful process. Wounds were deep, and the South was so badly crippled, there was almost nothing that could be salvaged. But that is a tale for another book. Abraham Lincoln would live to see the end of the war but not the rebuilding. However, before leaving the war behind, it is worth noting that the war was not the only thing Abraham Lincoln had on his plate during this era.

Lincoln's administration accomplished more than simply getting the nation through the Civil War. Lincoln signed into law a number of far-reaching acts that would transform the country, but most especially the "great interior region" that would become known as the Midwest. People had rushed to settle the better-watered and more easily farmed land east of the Mississippi, and enough of them had made it across the big river that Iowa and Kansas achieved statehood before the Civil War, but comparatively few had ventured out onto the Great Plains. In 1862, Lincoln approved the Homestead Act, hoping to encourage further westward expansion even with the Civil War in progress. This law offered 160 acres to anyone who wished to farm, with only a small registration fee and a promise to "improve" the land (that is, to farm it) for five years—so virtually free land. If, after six months, the settler decided he wanted to purchase the land outright, it was only $1.25 per acre. The only rules for obtaining land were that the potential homesteader be twenty-one years old and the head of a household, making land available to everyone who wanted it, including women and African Americans.[36]

Also signed into law in 1862 was the Morrill Act, or Land Grant College Act. This act granted each state thirty thousand acres of public land (or the equivalent in scrip, if land was not available) for each of the state's congressional seats. States were to sell the land or scrip in order to build colleges that would teach agriculture and mechanical arts, to meet the nation's need for agriculturalists and technicians who would be scientifically trained and forward thinking.[37] Sixty-nine land-grant schools were created as a result of this act, including many that would become the Midwest's top universities.[38]

The Morrill Act was actually not a new idea, but rather created a means of implementing ideas originally suggested by Thomas Jefferson. Its concept of combining liberal arts education with science and agriculture was remarkable. However, it was the democratization of education that made it truly a game-changer. It put a college degree within reach of all high school graduates who wished to pursue higher education. It offered education to the working classes, women, immigrants, and minorities. It was the beginning of great social change in the United States.[39]

The combined benefit of the Homestead Act and Morrill Land-Grant Act was the creation of a political and economic climate that would benefit everyone, including small farmers. The government had clearly acknowledged the importance of farming to the success of the country.[40]

In a way, these two laws, though created by Congress, were a culmination of the vision Lincoln had shared during an 1859 address at the Wisconsin State Fair. At that time, Lincoln had said, "Let us hope, rather, that by the best cultivation of the physical world, beneath and around us; and the intellectual and moral world within us, we shall secure an individual, social, and political prosperity and happiness, whose course shall be onward and upward, and which, while the earth endures, shall not pass away."[41]

Another key bit of legislation that Lincoln signed in 1862 was the Pacific Railway Act. This would further open up the interior of the land and tie the nation together, from coast to coast. Lincoln had a tremendous appreciation of railroads and what they could accomplish (no doubt influenced by the fact that he had, in his youth, driven herds of hogs to market on foot). The Pacific Railway Act would give land and government bonds to the Central Pacific and Union Pacific Railroads, with the goal of creating a transcontinental railroad. Lincoln would not live to see the project completed, but on May 10, 1869, the golden spike was driven into the railroad, uniting east and west. The trip to California from the East Coast had been reduced from six months to two weeks.[42] At the same time, that uniting marked the end of the frontier.

In 1862, Lincoln also signed into law the creation of the United States Department of Agriculture. In addition to the new methods of meat processing, canning had been invented in the early 1800s. With food being processed by unknown people at often-distant businesses, rather than at home, some consumers were concerned about the quality and safety of these foods. In addition, trains were now moving foods across state borders, so problems could quickly spread. The U.S. Department of Agriculture had the dual purpose of ensuring that every American involved in agriculture had access to information that would aid in all aspects of farming[43] and protecting the quality of food supplies, with a goal of developing standards for food processing. While the new government agency was created to help farmers, it would also aid in building the industry that would make food more readily available than it had ever been before.

So, the period of 1861 to 1865 changed everything. War divided the country as trains united it. The Midwest helped decide the war, and new laws helped develop the Midwest. But one thing remained constant. There were still a lot of pigs.

Getting Back to Growing

On December 25, 1865, eight months after the end of the Civil War, Chicago's business elite celebrated the opening of what would in time be hailed as the eighth wonder of the world: the Chicago Union Stock Yards.[44] Throughout the war, pigs had been processed at smaller yards all over Chicago, but this new effort would unite all those smaller yards in one vast, sprawling enterprise that would lead to Sandburg's famous definition of Chicago as being "Hog Butcher for the World." Indeed, this was no exaggeration. In the first week of operation, the Union Stock Yards processed 17,764 hogs. By 1900, Chicago processed 82 percent of all meat eaten in the United States.[45] By that time, "the yards" were also attracting roughly half a million visitors annually, who happily paid for tickets to tour the astonishing facility.[46] (If it seems surprising that people would pay to watch animals processed, it helps to remember that, at that point, almost everyone alive had either seen an animal butchered or had done it themselves. The attraction here was that it was done with stunning speed and efficiency. It was the epitome of the concept of "modern.")

The Chicago Union Stock Yards, along with its close associate, Packingtown (as in meatpacking), would do more than simply process livestock. They would change the world. Even with cities and farms separated, up to this point, animals were still herded into town and processed, usually in plain sight, by neighborhood butchers. Participation even at this level was now removed. The trains going directly to the Stock Yard solved the problem of animals strolling through town. No more dodging pigs or stepping over blood in front of the meat shop. It was all handled somewhere else. Then as now, animals raised on family farms across the Midwest became industrial products when they entered the modern facilities. On the plus side, it made meat both cheaper and more readily available.[47]

But making meat less expensive and separating the population from their food sources (for good or ill) were not the only changes to come out of Chicago. Ice harvesting—cutting blocks of ice out of lakes during winter and storing it for later use—became a huge business. It was needed for cooling

Photo 6.1. Meat went from anonymous to branded with the rise of the meatpackers. Photo by Cynthia Clampitt, taken at Jake's Country Meats, Cass County, Michigan. Used with permission.

both storage facilities and refrigerated trains, which, combined with expanding train service, meant the old tradition of slaughtering pigs in late fall or early winter could be abandoned. Meat could be had anytime.[48]

The yards held the animals—millions of them. Packingtown turned them into products. Names associated with Packingtown, such as Philip Armour and Gustavus Swift, would soon become familiar, as the concept of branding products developed.[49] By the 1880s, there were twenty-nine major packinghouses and several small ones in Packingtown. The scale of operations kept growing. By 1880, the Union Stock Yards were receiving more than seven million hogs annually. Employees numbered in the tens of thousands. Ice harvesting expanded further; Swift alone was using 450,000 tons of ice annually to chill meat. By 1900, the yards covered 475 acres and had 130 miles of train tracks along their perimeter. Unfortunately, pollution in the area also increased.[50]

The speed with which food was processed, not to mention the issues surrounding disposal of whatever parts of the animals could not be marketed, plus the newly "mysterious" process of producing meat by a comparatively small number of people, began to make the public wonder about their food. On top of this, President Theodore Roosevelt was suspicious of price fluctuations, assuming they must be corporate manipulations rather than simply good or bad crop years. He ordered James Garfield, son of the former president, to investigate. When Garfield's massive report turned up the fact that price fluctuations were, in fact, tied to crop fluctuations, Roosevelt ordered another investigation. The publication in 1906 of Upton Sinclair's book *The Jungle* gave Roosevelt what he really needed: public attention and concern, though for reasons different from those Roosevelt had in mind. (Interestingly, Sinclair was distressed that people thought his book was about the animals. A socialist, he had meant it to be about workers' rights. He was stunned to find out that people cared more for the pigs.) Though clearly a work of fiction, since nothing was as bad as what Sinclair wrote, *The Jungle* had created enough public response that Roosevelt had his justification for passing the Pure Food and Drug Act and Meat Inspecting Act in 1906.[51] Unfortunately, this didn't address consumers' biggest concern, which was trichinosis, but a large, centralized operation was an easier target.

The controversies about processing did not end meat consumption, but food fads would cut into the numbers. When Battle Creek, Michigan, health guru Dr. John Kellogg invented corn flakes, some people began switching to cold cereal for breakfast, rather than meat. Also, people in cities began to realize their desk jobs didn't require as large a food intake as working on a farm. Several decades before Dr. Kellogg tried to steer people away from meat, Edward Hitchcock, president of Amherst College, had warned that bacon should only be eaten by those doing heavy labor and should be avoided by "the sedentary and the literary." Then there were the cutbacks caused by the Meatless Mondays and Porkless Days of World War I, followed by the privations of the Great Depression. But these still only represented a relatively modest decline in American meat consumption. In fact, during the Great Depression, the government distributed salt pork to those in need, and even published a cookbook of salt pork recipes. Americans viewed abundant, cheap meat as a right, and they weren't likely to give it up just because of a food fad, war, or shaky economy.[52]

By the mid-1900s, diesel trucks and the expanding highway system meant work was no longer tied to railroads. The big processing companies relocated from Chicago to Kansas City, Missouri; Omaha, Nebraska; East Saint Louis, Illinois; St. Joseph, Michigan; and Green Bay, Wisconsin. The Chicago Union Stock Yards, which had for decades never processed fewer than thirteen million animals a year, began to decline, and in 1971, were finally closed.[53]

In the late 1800s, while Chicago still dominated the meatpacking scene, the rest of the region, especially west of Illinois, was still growing and developing. By 1880, Iowa passed Illinois for numbers of swine, with more than six million, compared to five million in Illinois. While numbers in both states would continue to grow, Iowa would grow faster and lengthen its lead, with nearly ten million swine by 1900.[54]

Nebraska had become a state in 1867, though South Dakota and North Dakota didn't achieve statehood until 1889. But finally, the "Greater Midwest" was complete. As was true of previous Midwestern states, these states also focused on agriculture. In 1883, Omaha, Nebraska, opened its own Union Stock Yards. While the Omaha facility never got as large as Chicago's yards at their peak, in 1955, as Chicago declined, Omaha pulled into the lead. For a few years, Omaha was home to the largest livestock market and meatpacking center in the world. But, as with Chicago, its days were numbered, and by the late 1960s, it, too, closed.[55]

As was true before the Civil War, so too afterward, companies that are not in any way associated with pigs today often got their starts in association with pigs. The first step in processing a pig carcass is to scald it, which loosens the bristles and makes them easy to scrape off. In 1883, John Michael Kohler, who owned a company in Wisconsin that made cast-iron farming implements and ornamental hitching posts, devised a process of covering metal with enamel. He used this process on a large basin intended for scalding hogs or watering horses, but Kohler also thought that it had potential for indoor use. So, when the next catalog came out from the company, it noted that, when furnished with four legs, the hog scalder could serve as a bathtub. Kohler Company soon diverged from farming equipment into the indoor plumbing fixtures for which they are now known.[56]

Mortgage Lifters

Even as beef became more popular in some areas, especially with the rise of the western cow towns and railheads bringing cattle north from Texas, pigs were still a popular choice for Midwestern farmers. Whatever else people raised, they always had a few pigs. The 1903 *Iowa Yearbook of Agriculture* reported, "The hog is the poor man's friend. He will respond quickly to good care, will make money quickly and fast if of the proper type and given the proper care."[57] Having pigs could make the difference between succeeding and going bust. As a result, in the 1870s through the Great Depression, pigs were known as "mortgage lifters," because they were such a certain source of income. If nothing else sold, one knew that payments could still be made.[58]

And so, the Midwest became the Corn Belt, and most of the Corn Belt became the Hog Belt. Aside from the abundance of corn, the greatest advantage offered by the region was wide, open spaces. It made it possible to keep pigs far away from most consumers. In fact, hogs today are raised on average 150 miles from any major population centers.[59] So unlike the 1800s, when pigs were still found in many city streets, one can now live an entire lifetime without ever seeing a pig. Which is actually kind of a pity.

There would be more food fads (food fads actually only being possible in areas with great abundance), food panic (or, more precisely, fat panic), new breeds and goals for breeding, ad campaigns, research and development, and ways of consuming pigs, but there would always be pigs.

SECTION II
PIGS ON OUR PLATES

CHAPTER SEVEN

Versatile Pig

The Parts and How We Use Them

One facet of pigs that has captivated people pretty much from the beginning is that each creature supplies a tremendous variety of tastes and textures. In the first century A.D., Roman scholar Pliny the Elder rhapsodized in his encyclopedic *Natural History* that "there is no animal that affords a greater variety to the palate of the epicure; all the others have their own peculiar flavour, but the flesh of the hog has nearly fifty different flavours."[1] And more than a century before Homer Simpson doubted the existence of the "wonderful, magical animal" that could provide such varied delights as the pig does, British author Isabella Beeton noted that "no other animal yields man so *many* kinds and varieties of luxurious food as is supplied to him by the flesh of the hog differently prepared."[2]

Indeed, the pig offers a wide range of options. However, the parts most valued and the options most relied upon have varied with era, culture, economies, tradition, class, market demand, and available technology. We no longer swoon over pig udders, as they did in ancient Rome, but even within the realm of parts for which there is still a demand, definitions or methods have changed. For centuries, after the hams were removed, most of the pork was simply cut into large chunks, barreled, and brined. Better cuts might be identified and put in barrels with higher price tags, versus the lesser cuts, which often went to sea with the world's navies. Odd bits and trimmings might go into sausages. But on the whole, other than when the pig was first slaughtered, fresh meat was rare. Pork was rarely sold as identifiable "cuts"—with the exception of ham, which even in prehistory was in a different category than the rest of the pig.

Of course, one need not always identify a favorite part. A number of cooking options utilize the entire animal, such as whole-hog barbecue or having a pig roast for a party. But other than for restaurants, competitions, or festive occasions, most of the contact we have with pigs today is in parts.

The cuts described below are the most common cuts in the United States, along with some more recently introduced cuts. Large cuts that can be broken down further are called *primals*. For example, the loin is a primal, or initial cut, which can then be broken down into roasts or chops. The pork cuts illustration shows where the primals are located on the pig's body, as well as a few of the "other" parts, such as ears and trotters.

Because it has historically had the greatest snob appeal, we'll start with ham.

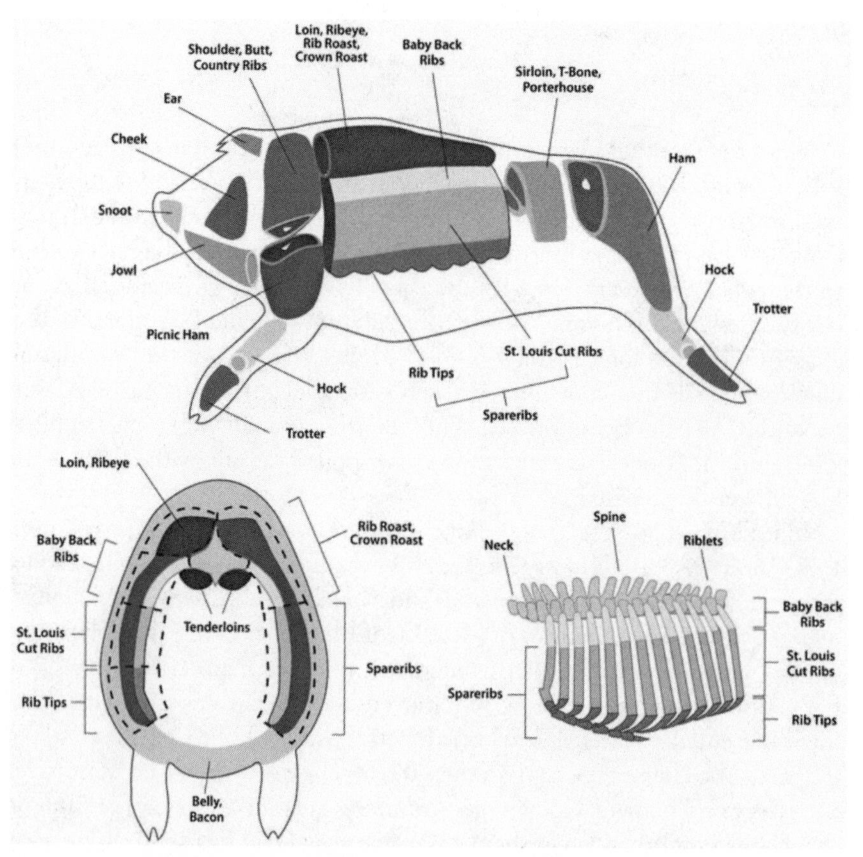

Photo 7.1. This chart shows the locations of the most common cuts of pork—and makes it clear that loin does not equal tenderloin. Chart courtesy of Craig "Meathead" Goldwyn of AmazingRibs.com.

Hams

Hams were and are the aristocrats of pork products. Because of their heroic size and handsome appearance, they escaped the image of "food for the poor" that clung to the rest of the beast. A ham is a rear leg of a pig, above the hock. It is the largest muscle on the animal, with the two hams together making up roughly 18–20 percent of a pig's weight.

The curing of hams dates back roughly four thousand years, to the time when Celts were still primarily salt miners in what is now Austria. The Romans imported Celtic hams, but the writing of Roman senator and historian Cato the Elder suggests that some Romans learned to cure and smoke their own. In his treatise *De agri cultura* (On Farming), written about 160 B.C., Cato relates a method for preserving hams that would have seemed familiar to most people up until about a century ago.

When thinking of ham today, Americans tend to think of cured ham. Given the ham's history, that's not unreasonable. However, ham is a specific muscle, not how it's prepared, and one can find fresh ham—though often, to avoid confusion, it will be referred to as fresh pork leg.[3] However, while recipes for fresh ham abound, most Americans eating fresh ham are enjoying a whole roasted or barbecued hog.

In the United States, we generally divide the cured-ham world between country ham and city ham. While curing for both types starts with salt, country ham is dry cured and aged, and city ham is wet, or brine, cured, and not aged. Country ham is the older of the two, tracing its lineage to Europe, and was for much of history the only type of cured ham.

Among early American dry-cured hams, the ones from Smithfield, Virginia, made from peanut-fed pigs, were the first to attain celebrity status. Queen Victoria had a standing order for six Smithfield hams a week. In 1926, the Virginia General Assembly passed a statute defining the process of making a Smithfield ham and declaring that to be called a Smithfield ham, the meat has to have been processed in the town of Smithfield.[4]

The wet-curing process for city hams is not very different from the preparation of barreled or pickled pork—the change was doing it to ham. Americans had dined on wet-cured, barreled pork for a few centuries, so the new city hams did not really offer an unfamiliar taste or texture.[5] Using a brine makes salt penetrate more quickly, and once meat processing started to become big business, rather than a family or small-town task, "more quickly" was a key qualification for adoption of a technique. As the twentieth century progressed, most meatpackers switched production from country hams to city hams.[6] In all fairness to the meatpackers, this change was not just for their convenience. City hams required little preparation before cooking, and some even came precooked. Country hams, on the other hand, while offering richer flavor and longer shelf life, required a considerable amount of work on the part of the cook. Drier and saltier than their city cousins, country hams must be soaked for anywhere from twelve to thirty-six hours, depending on the ham, with the water being changed every four hours. After that, the ham can be cooked.[7] As the pace of life increased during the twentieth century, people were happy with the ease of city hams.

The next big leap into the modern world occurred in Austin, Minnesota, in 1926. George A. Hormel had transitioned from small-town meat market to corporation in 1891, with a dream of competing with the big packinghouses in Chicago. Hormel worked alongside his employees, trimming hams and bacon, to make certain the job was done right. It was said of him that he "looks over his ham the way most men look over a pretty girl." Cured hams lasted longer than fresh, but Hormel wanted them to last even longer. For three years, the company had experimented with canning technology, never managing to create an adequately efficient process. Then, in 1926, Hormel invited expert canner Paul Joern to travel from Hamburg, Germany, to Austin, Minnesota. Within four months, with Joern's expertise guiding them, Hormel introduced his Flavor-Sealed Ham—the world's first canned ham.[8] Commercial canning had actually emerged in the mid-1800s, and sardines, oysters, and chopped meat were among the first items canned, since they are so perishable. These were, however, small cans. Canning an entire ham was a dramatic innovation, and it was an immediate success. Other companies followed suit, and today there are numerous brands of canned ham worldwide.

Through the twentieth century, processing methods continued to be altered in the hope of making production faster, thus lowering prices further. However, there was always a tension between increasing speed and protecting flavor. Because it takes time to develop flavor in a ham, there was a degree to which a fast ham would never have the flavor of a ham aged for a year or two. Still, hams retained their status as food for celebrations.[9]

While city hams are now more common in American life, country hams still exist and have begun to garner more attention of late, as people have become more interested in reviving tasty traditions. In addition, there are those who are turning hams into delicacies once limited to the Old World—that is, making prosciutto. Today, we have more options than ever before, with ham packaged in myriad forms—or not packaged at all, from local butchers.

Bacon

If the only image the word *bacon* conjures up for a potential diner is fatty, smoky, sizzling strips of pork belly, that diner is probably an American. As Samuel Johnson defined it in 1755, bacon is "the flesh of a hog salted and dried." So, the term *bacon* applied to any piece of pork that was cured.

Actually, if one goes back even further than Johnson's definition, *bacon* originally just meant "pork"—fresh or cured. That's why what we now know as meat pigs were originally called bacon pigs. Then, in the 1300s in England, the definition was narrowed to being the side and back. In fact, the word *bacon* comes from an old German word that meant "back." It was in England that bacon as we now know it originated, though included with belly bacon in that "as we now know it" are both English bacon, which comes from the back, and Canadian bacon, which is the loin.[10]

In early American history, that same definition held—until meatpackers started tinkering with the concept. In the late 1800s, bacon and ham were the most popular of the cured meats, though ham was popular for celebrations and bacon was everyday food. Wanting to take advantage of this, and also wanting to find ways to brand meats, meatpackers began to look for new approaches to processing bacon. Cincinnati and then Chicago began to define bacon as cured, smoked pork bellies. Dry curing offered more intense flavor, but wet curing gave more consistent results and kept prices down. And while discussions of what really constituted bacon persisted for a while, by the early 1900s in the United States, it was definitely pork belly. Before World War I, bacon was sold in slabs of four to ten pounds. Then, by 1915, a few producers began to offer presliced bacon. By the mid-1900s, while slab bacon was still in butcher shops, pretty much all commercial bacon was sold sliced. The process for creating bacon became so automated that everyone could afford it.[11] Among the many sources that reflect this widespread availability is Steinbeck's *The Grapes of Wrath*. Set during the Great Depression, the novel's main character, Tom Joad, comes upon another itinerant worker frying bacon, and when offered some, his response is, "Smells so darn good I couln' say no." To this, Joad's host replied, "Ever smell anything so good in ya life?" So, thanks to new processing, even the poor could afford bacon.

With pork production settled in Chicago and the debate over what constituted bacon beginning to be resolved, meatpackers could focus on other business issues. With bacon reduced to a specific cut, it could be more easily branded. Barrel pork was anonymous, but bacon now had names such as Swift and Armour. Once slicing bacon began, it was even easier to package and brand, with packaging done by hand and the wrapping mimicking the white paper used by butchers.[12] Then, in 1924, Oscar Mayer patented a machine that could wrap the sliced bacon.[13] It was hoped that packaging bacon nicely would elevate its reputation, which was largely that it was for poorer people.[14]

The next innovation in the rush to make bacon more readily available was making it easier and faster to slice. In 1930, machines could produce three hundred slices per minute, but the pork bellies had to be hand trimmed to fit the slicing machines and then hand fed into the slicers. A belly roller made the bellies all the same thickness, and then a device called a belly press was developed. The belly press turned every pork belly into a shape and size that was perfect for the slicing machines. This really sped up the process. By 1952, the belly press had been fine-tuned to where it could handle 240 bellies per hour, and speeds just kept going up from there. (High-end bacon producers generally eschew the use of the belly press, because it alters the meat, but, as is usually the case, finer quality means a higher price.)

Just as ground spices lose flavor faster than whole spices, so, too, sliced bacon lost flavor faster than slab bacon. So, packaging had to advance, as well. Packinghouses started using cellophane in 1920, which helped slow the deterioration of flavor in the sliced bacon. Vacuum packing was introduced in the late 1950s.

By the 1960s, statistics were reflecting the success of the meatpackers. The price of bacon had dropped dramatically, and the acceptance of bacon across all economic and social levels was close to universal. Bacon in the United States had been transformed.[15]

And then, in 1961, it was transformed again. It became a commodity. It was in Chicago that pork bellies began being traded by the Chicago Mercantile Exchange. Pork bellies were the one trade item that even people with no knowledge of commodities or futures trading could name. In a time when bacon was largely seasonal (summertime bacon, lettuce, and tomato sandwiches being the primary use), buying pork bellies, freezing them, and then selling them when demand increased was a perfect way to even out the trading cycle. But then bacon became even more popular. Bacon was selling year-round. This success derailed the pork belly market, and in 2011, the Chicago Mercantile Exchange ceased trading pork-belly futures.[16]

There are a few other applications besides bacon for pork bellies. In parts of the Midwest, one comes across side pork, which is pork belly that has not been smoked or cured. In other words, it's fresh pork. It is sufficiently popular in Wisconsin to warrant a Side Pork Festival, held annually in Sheboygan. If side pork is salted, it becomes salt pork—at least today it does. Historically, salt pork could be any part of the pig, and even the whole pig, but if one goes to the store today to buy salt pork, one will find only salted pork belly. So, salt pork lives on, which is good, as there are still so many recipes that demand it. And pork belly has multiple forms, with bacon, side pork, and salt pork, plus the increasing popularity of crispy, fatty, tasty dishes that simply utilize pork belly.

Perhaps worth mentioning before moving on to other cuts is the announcement in 2015 by the World Health Organization that bacon (along with all other processed meats) causes cancer. The original statement by the World Health Organization was revised a month later, clarifying that only large amounts of bacon affect some people, but that overall, bacon is not a serious threat to health, and the International Agency of Cancer Research is not recommending that people stop eating it. Of course, the amount of fat and salt in bacon have long been viewed as indications that moderation is wise, but now, official health organizations are confirming that wisdom.[17]

Another ongoing worry about all processed meats, including bacon, is the use of nitrates and nitrites. Interestingly, most of our nitrate consumption (80 percent) comes from vegetables, and our bodies produce nitrites from nitrates. Nitrates and nitrites are actually necessary to health. Nitrates are even used as supplements for lowering blood pressure and enhancing exercise. In meat, they stop the growth of bacteria, making the meat safer, and improve the color. That said, as with so many things, it is the dose that makes the poison. One can get too much of a good thing. Nitrites can, in large enough quantities and in some people, be converted into nitrosamines, which are carcinogenic—though a bit of vitamin C keeps this conversion from occurring. Research will continue, but scientists now say that consuming reasonable quantities of processed meat, and making sure you have a balanced diet, ensures that you won't get more nitrates and nitrites than your body can handle.[18]

Tenderloin, Loin, and Chops

Before the advent of refrigeration, enjoying these cuts fresh was very seasonal. One might have pork chops, a hefty pork stew, or a nice roast the day or, depending on the weather, the week the hog was butchered, but for most of history, these cuts would simply have been salted or pickled, along with any other meat that didn't go to the smokehouse.[19]

Interestingly, it was those most likely to be doing the actual slaughtering who even wanted fresh pork. Middle- and upper-class Americans, particularly those living in cities in the northeast and Midwest, were strongly disinclined to eat fresh or uncured pork, a dislike that lasted well into the twentieth century.[20] So these cuts were, for a long time, the domain of small farmers and poor people, particularly, for many years, African Americans.

Loins, chops, and tenderloins all come from roughly the same area—the pig's back. A pork chop is simply a slice of pork loin with a rib still attached. With ribs removed, pork loin is a nice roast—though it is moister with bones attached, and a crown roast (where the loin and ribs are bent into a circle) was in the mid-twentieth century among the most impressive of presentation pieces for an elegant party.

Because a roast takes longer to cook, and, if one is not feeding a crowd, is more of an investment than many might wish to make, the pork chop is a great option. It offers the lovely, delicate flavor of the loin, but in a size that is more quickly cooked and more manageable to eat.

The tenderloin is often confused with the loin, but they are not the same cut. The tenderloin is so called because it gets almost no exercise at all, which means (not just in the pig, but in all meat) that it will be really tender, but not particularly flavorful. A pig has two tenderloins, each of which will weigh roughly one pound. The tenderloin is lean, delicate, and quick to cook, so perfect for a weeknight meal. It can easily be pounded flat, to create a schnitzel-like cut of pork, in which form it frequently appears across the Midwest. However, it is not a good choice for applications that involve longer cooking.

Ribs

Pigs can have anywhere from thirteen to seventeen pairs of ribs. While variations in number are hardly surprising among different breeds, with pigs, it's possible for differences to occur even within a single litter. In addition, with some regularity, pigs have been found to have a different number of ribs on one side versus the other. Gender does not affect the average number of ribs.[21]

Variations in numbers have no impact on the taste, or on the nomenclature, when trying to identify what type of rib one is ordering for dinner. Craig "Meathead" Goldwyn is a barbecue expert and the creator of the popular AmazingRibs.com website. He relates that, in the United States, ribs generally get divided into four categories: baby back ribs, spareribs, St. Louis cut ribs, and rib tips. Probably the easiest way to understand where these ribs come from is to refer to the pork cuts chart Goldwyn has shared, which appears at the beginning of this chapter. There is generally more meat on the baby back ribs, as they are the ribs attached to the loin muscle—and, therefore, are also the bones in bone-in chops. (Interestingly, while the loin was once the pricier option, the demand for baby back ribs has caused these ribs to pass pork loin in price.) Baby backs are the leanest ribs. Ribs have more fat as one gets farther from the spine and closer to the belly. The fat content, bone size, and length of the ribs selected will affect the cooking time, as well as the taste and texture of the ribs.

While valued by some groups of Europeans (it is hard to imagine Germans without their ribs, sauerkraut, and potatoes), in the 1800s, ribs were not highly regarded by many folks, particularly once people got to a point economically where they weren't forced to eat what they considered less desirable pieces. There was a demand, but the demand was so low compared to the demand for other cuts that, in Cincinnati during its pork-processing heyday, pork processors often threw the ribs in the river. The arrival of German immigrants in the 1840s began to change this. Spare ribs and other "throwaway" parts, such as trotters, were collected by the basketful by thrifty, pork-loving Germans, who loved the fact that their new home afforded them so much free food.[22]

One might think that throwing out the ribs was how they got the name "spare," but in fact, the word appears to have come from Low German *ribbesper*, which is pickled pork ribs roasted on a spear or spit.[23] Given that it was the Germans who most appreciated the ribs, this derivation makes sense.

Still, spare ribs were often "spare," in the sense of being something one didn't need but could give to someone else. When the person who was giving away the spare ribs was an individual, rather than a huge meatpacker, the meat was most often given to those who were viewed as needy. With so many immigrants arriving in the United States penniless, there was no end of people who qualified. In *Come and Get It*, food historian Robert Dirks describes the extent of poverty in one Midwest town and relates the efforts of one woman to help, including giving ribs to her poorer neighbors every time she slaughtered a pig.[24] This kind of generosity was common in the mid-1800s, as those with money were often at most only one generation away from having arrived in the United States themselves.

The same phenomenon occurred in the South as well, with the poor to whom ribs were given often being African Americans. Because it was African Americans who were so commonly responsible for cooking and, to an even larger degree, preparing barbecue, it is perhaps not surprising that barbecue ribs appear to have been perfected among African Americans in the South. It was the twentieth century's Great Migration that brought barbecue ribs to northern cities—Southern barbecue, that is, not Chicago's boiled ribs with sauce that locals identify as barbecue. As poet LeRoi Jones wrote in 1966, describing dishes introduced by African Americans into Harlem, "Ditto, barbecued ribs—also straight out of the South with the West Indians, i.e., Africans from farther south in the West, having developed the best sauce."[25]

With an origin that might best be described as an evolution, barbecue ribs have become part of Americana, adopted wholeheartedly across the nation.[26] They have been so completely absorbed into American culture that almost everyone could identify with the M*A*S*H episode in which Hawkeye decides he has to have barbecue ribs, no matter how hard it was to get them. Fortunately, because of their increased popularity today, it is far easier to come by good barbecue ribs than at any other time in history.

Newly Fashionable Cuts

As the snout-to-tail movement has gained strength, there has been a resurgence of interest in a more varied range of pork dishes, old and new. Rather than ordering specific cuts, or even primals, some restaurants have begun ordering whole hogs, which are then butchered in-house, to see just what the possibilities are. As part of this revived interest in "other" parts of the pig, diners have found new cuts appearing on menus and in butchers' cases, such as *secreto*, *porchetta*, and pig wings.

There is also a new focus on often-overlooked parts of the pig, such as the shoulder. While always seen as something of a poorer relative of the ham, since it is fattier and tougher, the shoulder is also very flavorful. The entire primal is traditionally called a Boston butt, but the lower half is sometimes called a picnic ham. Master butcher Kari Underly talks about the recent increased interest in the shoulder.

> A number of overlooked cuts can be created by taking apart a traditional Boston butt or the picnic. The butcher can isolate the picnic (triceps muscle), which makes a nice roast, or smoke and slice it for sandwiches. Sure beats using it for pulled pork. Grillable cuts from the Boston butt are the *coppa*, *pluma*, *presa*, and *secreto*. These items are being given greater attention because chefs must sell every bit at a higher price to make the financials work. This is the challenge with whole-animal butchery—making money.

Because eating everything isn't new, these cuts do have traditional applications. *Coppa*, also called *capocollo*, depending on the part of Italy in which it is found, is an Italian cured meat. *Pluma* and *presa* are cuts from near the neck that are popular in Spain.

Secreto is one of the tasty cuts that seem to have found traction in some pork-centric restaurants. *Secreto* is Spanish for "secret," and sometimes the secret is what is actually being served. Underly explains,

> This cut originated with Iberico Spanish hogs, hogs that are fed on acorns and have wonderfully marbled fat. Cuts like the short rib muscle from the shoulder, the bottom sirloin flap, and skirt steak are all items that express this flavorful marbling exceptionally well, and all these cuts have been known to be called the secreto. In my opinion, the true secreto is the short rib muscle, also known as the *serratus ventralis*. It rests next to the rib and under the scapula bone.

Porchetta is a handsome and impressive roast, the antiquity of which is confirmed by the fact that there are a few different versions, including one in Minnesota (called porketta and made from the shoulder) that is nothing like the one found in Chicago, which is closer to the Italian original, but not quite as complex. (The original Italian version, which was first developed in the 1400s, involved a whole, deboned pig cooked in a pit.)[27] However, the version that is becoming popular in high-end urban restaurants, while not an entire pig, is still impressive. Underley relates,

> *Porchetta*, which is an Italian dish in which the skin, belly, and loin are all used, has seen a resurgence in popularity due to the availability of whole-hog butchery. The skin and belly are rolled around the loin to protect it. It can also be stuffed with sausage, well spiced, or just add salt and pepper. The true treat is when the chef slow roasts this to perfection. The skin turns into cracklings, the pork belly melts into the loin, and this delicacy warms everyone's heart!

"Pig wings" are becoming popular bar food, appetizers, and snacks. The "wings" are made by cutting the fibula, the thinner second bone in the leg of a pig, away from the larger tibia, the heavy bone that we associate with the term "hock" or "shank." These so-called wings are small, with a nice mouthful of lean meat attached. Because they are being marketed in a niche similar to that of chicken wings, the name "pig wings" was adopted as sounding more appetizing than either "fibula" or "shank."[28]

Underly notes that it seems likely there will be a continuing increase in interest in pork as restaurants and their clients continue to pursue the farm-to-table and snout-to-tail concepts.

Sausage

In an article titled "Hot Dogs," published in 1929, journalist and cultural critic H. L. Mencken urged people devoted to American frankfurters to pursue instead the "vast and brilliant domain of the German sausage." He then explained why. "For there are more different sausages in Germany than there are breakfast foods in America, and if there is a bad one among them then I have never heard of it."[29]

Of course, while German-speaking peoples do have an astonishing array of sausages, the fact that Mencken was of German descent probably influenced his specifying only German sausages. Because pretty much every culture associated culinarily with pigs has some form of sausage, and usually multiple forms. Sausages can be fresh or dried, with or without casings, made from any number of parts of the pig, including (and this is unique among animals used in sausage making) the blood. While sausage is generally easily recognizable as sausage, no matter where one finds it, this food form varies tremendously from country to country, and even region to region. The reason for the ubiquity of sausage is, of course, that it makes it possible to use everything. If dried and smoked, it also has the benefit of lasting for a long time. And it is generally exceptionally tasty.

The invention of sausage occurred roughly five thousand years ago, with the Assyrians often credited with the first sausage. That said, it's very difficult to nail down any development that is that far back in history. Fortunately, the idea took hold, spreading outward as people traded and conquered and migrated.

Sausage is essentially ground meat combined with fat, salt, seasonings, and sometimes starchy fillers, as with scrapple. We tend to think of it as being stuffed into a casing, but there are many sausages that are not (think of bulk sausage, sausage on pizza, and sausage patties at breakfast). While sausages can be made from almost any meat, pork has long been the top choice simply because fat is so vital to successful sausage making, and fat is something one can rely on with a pig. Sausage turns humble scraps into a luxurious feast that is at once remarkable and affordable.[30]

While casings are not required for sausage to be sausage, they are common. Natural casings are, of course, intestines, large or small, from whichever animal offers the right size for a specific sausage. Then, in 1925, Erwin O. Freund began production of the first synthetic sausage casings. He established his company, the Visking Corporation, in the best possible place for such a product: the Chicago Union Stock Yards. The casings were made of cellulose, and they offered the advantages of not needing to be thoroughly cleaned beforehand, as natural casings had to be, plus size could be absolutely consistent. Freund also discovered that some sausages, once cooked and smoked, could have their casings removed, and so the first "skinless sausages" were born—a most important development for hot dogs. The development of these synthetic casings revolutionized sausage manufacturing, making it faster, easier, and more consistent.[31] Natural casings did not vanish, and there is a resurgence of interest in natural casing these days, as many value the "snap" that they offer when bitten, and many traditionalists feel that it is something of a cheat to use cellulose. But Freund's invention contributed further to making sausage affordable.

While some sausages include the word sausage in the name (summer sausage, breakfast sausage), most don't. Salami, kielbasa, bologna, boudin, wiener, andouille, chorizo, linguiça, lap cheong, scrapple, olive loaf, anything that ends in *wurst*, and innumerable others fall into the wide embrace of the term *sausage*.

It is easy to settle into buying the big-name brands of sausage—and there is no reason not to, as the quality is good. However, there is adventure to be had if one wishes to look for it. Local butcher shops or less-familiar brands found at grocery stores can offer tastes of other lands. Online food sites and food magazines are filled with "ten best" lists of best sausage, best sausage maker, best place to buy sausage, so it need not be an unguided tour. But the world of sausage is one of astonishing variety, if one chooses to explore.

Photo 7.2. This display at a suburban Chicago grocery store underscores the variety of traditions and preparations: Italian, Russian, Hungarian, German, prosciutto, salami, pork loin, bacon, mortadella, and more. Photo by Cynthia Clampitt at Garden Fresh Market. Used with permission.

Lard

Rendering, or melting down, the fat from a pig's carcass produces lard, which is valued by cooks today but was absolutely vital for much of the world's history. However, not all lard is created equal. The *flare*, which is the fat around the kidneys and inside the loin, produces leaf lard when rendered. Leaf lard is the lard that pastry cooks seek out. The next level down is the lard produced from back fat, which is the hard fat layer between the skin and the meat. This is a good-quality lard that can be used for baking and frying, and was even spread on toast in place of butter, which was something of a luxury, either because it needed refrigeration or, as was often the case for farmers a century ago, it was simply too valuable to eat, rather than sell.[32] (That said, there are many countries that still consider lard in some form a delicacy, such as the *lardo* of Italy or the *smalec* of Poland.)

The caul is the fatty membrane that lines the pig's abdominal cavity. It is thin and rather lacy looking. It is used to wrap delicate dishes such as pâtés, forcemeats, and the like. During cooking, the caul fat melts into the food it has enclosed, adding to the flavor and keeping it moist. Unlike simple lard, one generally needs to go to a good butcher, rather than the grocery store, to get caul fat.[33]

In the 1700s and 1800s, lard was sometimes more valuable than meat. When the demand was high, packers would dump whole hogs into the rendering vats, sacrificing the meat to get more fat. However, lard became much less valuable once John D. Rockefeller established oil drilling and processing as an industry, following the Civil War, and petroleum products took over for oiling machinery and fueling lamps. Also, vegetable oils were created or became more widely available. Then, after World War II, the panic about animal fat being bad led to a rush toward alternative fat sources—or to eliminating fat. Pigs were bred leaner, not just because people were worried about fat, but also because the government was pushing for it. The U.S. Department of Agriculture developed information on genetic lines of the leanest pigs. Meatpackers got on board with the program, too, and Hormel Foods even awarded prizes to farmers who raised the leanest pigs. By the 1970s, the fat yielded by a pig had been cut by nearly 40 percent.[34]

Of course, we now know that animal fat is not evil. The bonus with lard is that it has less cholesterol and more unsaturated fat than butter. There are cuisines and cultures that have never backed away from lard, because it is such a good cooking medium and versatile ingredient, but today, an increasing number of those who once shied away from it are discovering its uses. It is great for frying. There is nothing like leaf lard for perfect, tender, flaky piecrusts. It's wonderful for roasting vegetables.[35] However, know that the lard generally found on the shelves at the average grocery store will most likely have been made from a blend of all the fats available from the pig. If one wants leaf lard, one has to look specifically for leaf lard.

Everything Else

When thinking of nose-to-tail cooking, it is hard to imagine anything that offers more options for practicing this concept than the pig. In addition to all the well-known cuts, there has been a resurgence of interest in ears, snouts, tails, hocks, and trotters. All these parts have been used in the past, but after decades of relying on hams, bacon, and pork loin, Americans are becoming more interested in what has been left behind. There are traditional French recipes for *pieds de porc* (trotters/pigs' feet), but in the United States, the most common treatment in the last couple of centuries has been pickling. They were so popular in Cincinnati in the 1800s, pickled pigs' feet became known as "Cincinnati oysters."[36] Today, pigs' feet have returned to high-end restaurants.

A few "other parts" have never gone out of fashion. Pigs' liver in the form of liver sausage (aka liverwurst) and close cousin braunschweiger (which is smoked and may add bacon to the liver) are Midwest staples. Pork skin never faded from the scene, but it has exploded in popularity among snackers who are avoiding carbohydrates. While chicken skin or duck skin can be a crispy delight, among mammals, the pig is the only one that has skin that is not simply edible, but pursued. We call the skin pork rinds, cracklins, chicharrón, or bacon rinds, and these crunchy snacks can be purchased almost anywhere. Stores with Hispanic or African American clienteles might seem like obvious places to look, but pork rinds can be found in the snack aisles of most stores. And while the big snack-food companies offer pork rinds in their product lines, Ohio-based Rudolph Foods, founded in 1955, doesn't make anything else besides pork rinds and cracklins.

Pig-ear sandwiches have been featured on *Diners, Drive-Ins and Dives*, though Michigan-based heritage breed pig farmer Nate Robinson says the best pig ears he's ever had are those prepared by Chicago chef Paul Kahan. Then there are pork sweetbreads. And intestines, either for casings or chitterlings. Pretty much everything gets used.

The teeth, bones, and bristles are not consumed, but they are utilized. The ability to sell byproducts has historically meant the difference between profit and loss in the meatpacking industry, so nothing is wasted. Bones become buttons or fertilizer. Bristles end up in hairbrushes. Gelatin, and even cortisone, can be derived from pigs.[37]

As the nineteenth-century lawyer and food critic Grimod de la Reynière wrote, "Nature in its perfection arranged it that everything in a pig is good and there is nothing to throw away."[38]

Variations on a Theme

While all pigs are *Sus scrofa domesticus*, not all are the same. Rediscovering the heritage breeds developed in the 1700s and 1800s is today's passion. This has resulted in a surprising alchemy: turning an animal that was traditionally the food of the poor into a magnificently exceptional dining experience that is economically out of range for the majority of the population.

England's love affair with pig breeding gave the United States the currently popular Berkshire. Interestingly, there was also something of a mania for Berkshire pigs in the 1840s.[39] Other popular heritage breeds from the British Isles include the Tamworth, Hampshire, and Gloucestershire Old Spot. Some of the best-known heritage breeds of American origin are the Duroc, Ossabaw Island (descended from pigs abandoned by the Spanish), and Mulefoot.[40] Not yet widely known, but still of great interest among pork connoisseurs, is the rare Mangalitsa, a Hungarian breed that is remarkable for its fatty, well-marbled meat (up to 60 percent fat!) and heavy, rough, wooly coat.[41] Just as people bred pigs for flavor and tenderness once trains entered the picture and pigs didn't have to walk hundreds of miles to market, so now some are looking to heritage breeds for traits other than the more traditional market-driven ideals of rapid growth or reduced fat.

At least until the next trend emerges.

CHAPTER EIGHT

American Icons

Barbecue, Hot Dogs, and SPAM®

As people spread across the United States, different regions developed different specialties, depending on the ethnic groups that settled the areas, as well as on available resources. However, there were a few pork-based food forms that arose that would become iconically identified with the United States and American culture, not just at home, but worldwide. Each appeared at a different time and reflected a different aspect of American history. All remain important today. These icons are barbecue, hot dogs, and SPAM.

Barbecue

There are disagreements in some quarters as to just exactly what constitutes barbecue, but no one disagrees that barbecue is quintessentially American. People have since time immemorial roasted animals over fire, but it was from the indigenous people of the Caribbean/West Indies that Europeans first gained the ideas of low, smoky fires and long, slow cooking methods that would evolve into classic American barbecue.

The Caribbean point of origin is reflected in Samuel Johnson's definition of *barbecue* in his *Dictionary of the English Language* (1755). He defined the verb *barbecue* as "a term used in the West Indies for dressing a hog whole; which being split to the backbone, is laid flat upon a large gridiron, raised about two foot above a charcoal fire, with which it is surrounded," but also included an entry for the noun *barbecue*: "A hog drest whole, in the West Indian manner." So, the concept of barbecue was clearly attributed to the Caribbean.

"In the West Indian manner" was "low and slow," which barbecue aficionados today have as their mantra. And, as Johnson's definition specified, the initial barbecues, once pigs were introduced, involved whole hogs.

The general consensus among scholars is that we got the term from the same place and people we got the cooking method. As mentioned earlier, the Carib called their greenwood grids *boucan*, while the Taino/Arawakan word for this cooking style was *barbakoa*, though some in the islands and in Brazil used the term *barbricot*. The Spanish settled on the word *barbacoa* and carried both term and method along the Gulf Coast, across Mexico, and into Texas.

The first barbecue in the British colonies was at Jamestown, Virginia, shortly after pigs were introduced. Barbecue spread as settlement did, though at that point, it was generally from Virginia southward. In 1663, King Charles II offered land to eight of his supporters, land that would become the Carolinas. Workers were solicited from Barbados, many of whom were Scottish and Irish men who had been indentured servants. Some of them brought with them the plantation system that had existed in the islands, as well as the reliance on slaves. All of them brought familiarity with barbecue and a reliance on pigs.[1] The barbecue that evolved along the southeastern coastal plain was generally open-pit style, with pigs cooked for eight to fourteen hours over coals made from local hardwoods. Chiles, either as part of the initial rub or part of a simple vinegar-based sauce, were another major New World contribution to barbecue—and tomatoes, indigenous to Mexico, would later get involved. But initially, it was mostly pork, salt, and lots of time.[2]

On May 27, 1769, a recently married George Washington wrote in his diary, "Went into Alexandria to a Barbecue and stayed all Night."[3] The diary reveals that Washington not only went to other barbecues, but also gave them, as did others with property and/or political ambitions. Barbecues had always been community affairs—one did not go to that much trouble to cook that much pork for just a few people. Barbecues celebrated homecomings. They were used to reward slaves for work well done. Churches would hold barbecues at which blacks and whites could mingle and worship together.[4] For politicians, barbecues became an important way to attract voters to rallies and, after the American Revolution, became the traditional way to celebrate independence. However, after the Revolution, barbecue began to fade from Virginian food culture, transferring its prominence to North Carolina.

Carolyn Wells, co-founder and executive director of the Kansas City Barbecue Society, says that North Carolina was the real "cradle of American barbecue." While some folks use the word *barbecue* to refer to anything cooked on an outdoor grill or baked and covered in bottled barbecue sauce, those who are serious about the tradition only apply the term to the slow-cooked barbecue that developed in the South.

Of course, part of the reason barbecues, especially the large-scale events, did well in the South was that slavery offered a readily available workforce to dig the open pits, cut the wood, and tend the coals over the many hours the meat cooked. As a result, barbecue most definitely became a major part of African American culture.

Three of the four major types of true barbecue are associated with Southern locations: North Carolina, Memphis, and Texas. The fourth location of the "big four" is in the Midwest: Kansas City. Carolyn Wells relates that

> Kansas City is the melting pot of barbecue. North Carolina is all pork and is cooked over hickory. Memphis is the pork barbecue capital of the world, and again you're using hickory. In Texas, beef is the protein of choice, and it's cooked over mesquite or post oak—so called because it's made from actual fence posts. In Kansas City, all three come together. Southern migration brought the pork traditions of the Carolinas and Memphis to us, and the railhead here brought beef cattle and barbecue traditions up from Texas. So, we have pork shoulder, pork ribs, and beef brisket in Kansas City competitions. Then, just so we'd have all the major proteins, we added chicken.

As an aside, using hickory and mesquite emphasizes the Americanness of barbecue. Both are indigenous to North America. The word *hickory* comes from the Virginia Algonquian word *pawcohiccora,* which means "food prepared from pounded nuts." Of course, one could use pecan wood, but that's a type of hickory, and the name is another Algonquian word, *pacan,* which means "nut." As for mesquite, that comes from the Nahuatl (Aztec) name for the tree: *mizquitl.* So, both the tradition and the fuel are anchored in the North American continent.

Barbecue is far from limited to the styles of the "big four." There are variations on the main themes all across the South. Some areas are defined by sauce, others by what they prefer to cook, and some by what they serve with the barbecue. South Carolinians fancy whole hog with a mustard-based sauce and consider Brunswick stew the appropriate side dish. In North Carolina, they like pulled pork, a vinegar-based sauce, and serve their 'cue with coleslaw. Mississippi, Tennessee, and Alabama will do whole hog, but their preferences run to shoulder and ribs.[5]

After the Civil War, barbecue continued to be important in the South, especially as it helped render the cheaper cuts tastier in areas devastated by war. When African Americans began to move from the rural South to the urban North, city churches used barbecues to help the newcomers adjust to their new surroundings, offering both comforting familiarity and a community that could help.[6]

That Great Migration northward would also create new markets for barbecue. Chicago was an important destination for African Americans moving north, not only because it's where all the trains went, but also because there were so many jobs with the railroads and in the Union Stock Yards. As a result, Chicago has developed a remarkably vibrant barbecue culture of its own. However, today, barbecue is no longer solely something one does for rallies or to find community (though those who do barbecue say that community remains a key part of the tradition). It has also become a fast-growing "sport." There are huge competitions in Memphis (Memphis in May) and Kansas City (American Royal Barbecue Contest), and smaller, Kansas City Barbecue Society-sanctioned events are multiplying at a remarkable rate. (More on this in the Popular Pig chapter.)

Barbecue anchors us in our history, in community, in less hurried times. Hot dogs, on the other hand, lead us into the industrial age.

Hot Dogs

In 1975, a television ad came out that placed hot dogs right in the middle of what the word "Americana" means: "Baseball, hot dogs, apple pie, and Chevrolet." Encased meats were nothing new. People had been stuffing chopped meat into innards for millennia. But the hot dog was different; it was a food that reflected what America was growing into.

The mid-1800s witnessed increased European immigration, rapid urbanization, and expanding mass entertainment. It was a time when the idea of being American was taking hold, adjusting to no longer being controlled by a European monarchy. The idea of being able to work hard and live well, regardless of background, was growing, too, along with a feeling of endless possibilities. Technology was viewed as the answer to problems, though it would prove in time to create other problems. But everyone was caught up in the rush of new things, and one of those new things was ways of processing meat.

Sausages had always been made by hand, by home cooks or small-town butchers, and they were made primarily from pork. In the United States after the Civil War, however, while pork would remain the most common sausage meat, the "made by hand" part of the equation began to vanish.[7] The age of industrialization was changing the speed at which everything was done, making a tremendous range of goods more widely available and less expensive than ever before. The Industrial Revolution was about a hundred years old when the Civil War ended, but it was just beginning its turn toward food. In

1868, a steam-powered chopper was introduced that could cut one hundred pounds of meat in half an hour.[8]

The new cutters would make the production of sausage easier, and, even more important to American consumers, cheaper. In addition to chopping faster, they chopped finer than was possible before, creating a smooth, pink meat slurry referred to as an emulsion or batter. This could be piped into sausage casings, cooked, smoked, and sold at bargain prices—just five cents for most of their early history.[9] These new sausages possessed a heritage anchored in the German-speaking world. They became known by the names of the towns from which the sausage makers and sausage traditions came: Frankfurt, Germany, and Vienna, Austria (Vienna being spelled *Wien* in German): frankfurters and wieners.[10]

These new sausages were fine-textured, more like bologna. More important still, they were precooked and a good size for eating on the go. So perfect for an increasingly bustling nation.

But why were wieners in buns "hot dogs"? In the 1890s, the word "hot" was being applied to lots of things, from hot music to hot women. A sharp dresser or top athlete might be called a hot dog. So "hot" and "hot dog" did not first appear in connection with wieners. While scholars are still searching for the earliest use of the term, one likely point of origin is a Yale University lunch wagon called The Kennel Club, also casually referred to as the Dog Wagon. The term spread first to college campuses and then to the general public.[11] "Echoes From the Lunch Wagon," which appeared in the Yale newspaper for October 5, 1895, seems to confirm this theory:

> "'Tis dogs' delight to bark and bite,"
> Thus does the adage run.
> But I delight to bite the dog
> When placed inside a bun.[12]

The public wanted cheap meat, and meat packers wanted something to do with the tons of trimmings their operations produced. (Trimmings are the meat and fat attached to an animal's skeleton that aren't part of major cuts of meat. It's what the home cook uses to make soup.) Using trimmings in these new sausages answered both needs—and they were such a success that more effort went into improving the machines. The trick became having enough workers to keep up with the choppers. In the 1900s, trimming departments had been developed, staffed entirely by women. Because of the importance of trimming, a good pork trimmer could earn thirty dollars per week, which was good money at the time.[13]

By the 1910s, choppers had been developed that could cut twenty-four hundred pounds of meat per hour. However, stuffing the sausage was still done by hand, as was the "linking," the separating of the long tubes of sausage into individual wieners. Men generally operated the stuffing machines, and women did the linking, deftly twisting the tubes at the required intervals.[14] Upton Sinclair described the sausage room in his novel *The Jungle*, writing that, as twenty- or thirty-foot-long tubes came out of the stuffers, a woman "seized them as fast as they appeared and twisted them into links . . . the woman worked so fast that the eye could literally not follow her, and there was only a mist of motion, and tangle after tangle of sausages appearing."

Those tangles of sausages might differ slightly from manufacturer to manufacturer, but short of having a chat with one's butcher, there was really no way to distinguish one from another. That is, until a Bavarian-born sausage maker in Chicago began focusing on ways to separate his products from the crowd. Oscar F. Mayer, who opened a butcher shop with his brother Gottfried in 1883, was a savvy marketer, and he wanted to make sure people knew that his wieners (and bacon and Thuringers) were of the highest quality and greatest purity possible. He worked to keep the company's name in the public eye, from sponsoring booths at the 1893 World's Fair to emblazoning the name on delivery wagons. Mayer's butcher shop graduated to full-fledged corporation in 1911 when Mayer's son, Oscar G. Mayer, graduated (with honors) from Harvard and returned to Chicago to help build up the family business.

In 1929, Oscar Mayer became the first company to actually brand its wieners. A yellow paper band, with the company's name in red, was placed around the sausages, so they could easily be spotted in retailers' meat cases. The yellow and red of that first branding device lives on in current packaging and has appeared in just about everything else related to Oscar Mayer wieners—including their next big marketing move: the introduction in 1936 of the Wienermobile. Sausage was no longer anonymous.

Hot dogs were "hot," but innovations such as the Wienermobile showed that companies still thought there was room for growth. In 1939, the World's Fair in New York City promised to give fairgoers a glimpse of the "World of Tomorrow." One aspect of that future vision was the exhibit created by the Swift Company, which actually put part of the process of wiener production on display—because high-speed processing was definitely viewed as a key to the world of tomorrow. Swift also made sure that anything that might be off-putting was out of view, with young, attractive women staffing the white-enameled production exhibit. Both fairgoers and the media were impressed.

Swift continued working on streamlining the process, but Oscar Mayer became the leader in this race, at least for a while. Oscar Mayer created an integrated system that they called the "wiener tunnel." It could produce ten thousand wieners an hour—not just cooked and finished, but wrapped in the equally new Saran Wrap™. By this time, World War II was over, and Americans were enjoying more hot dogs than ever. Baseball parks still did a booming business, but in the 1950s and 1960s, hot dogs began to move into people's homes. Then, in the early 1960s, a Des Moines, Iowa, engineer named Ray Townsend created the Frank-O-Matic, the first automatic sausage-linking machine. This machine could produce thirty-six thousand wieners per hour when it was first released, but continued improvements have raised that number to fifty-six thousand per hour today. Hot dogs were more abundant and cheaper than ever. Of course, there was a huge downside: all those thousands of skilled female sausage linkers were out of work.[15]

Because of its popularity in the United States, the hot dog became a symbol of America worldwide. The primary reason it became popular here is because it filled the need/demand for cheap, tasty, convenient protein, but the hot dog also reflected life. It was a thoroughly democratic food. In New York City, for example, construction workers, exchange students, tourists, and corporate executives rub elbows at the thousands of hot dog carts that dot the city. Hot dogs were the first "fast food," ideally suited for consumption at the baseball games with which they became so identified. They are also perfect for the suburban backyards that began to increase in number after World War II. Like Americans, they are industrial and urban, with the country's largest cities all being top consumers. Despite all the other changes in the world, America's love for the hot dog (in all its regional variations) remains constant.

SPAM

If there is one modern food that epitomizes the advantages that made pork so important throughout history, that food is SPAM. All the millennia of pork being affordable, readily available, and ideal for long-term storage are condensed into that little blue can. SPAM (or, more precisely, SPAM®) came officially into existence on May 11, 1937, when Jay Hormel, son of Hormel Foods founder George Hormel, registered the trademark. George had canned the first ham about a decade earlier, but Jay felt that, while ham was more upscale, pork shoulder was also worthwhile and, with the Great Depression snapping at the heels of most consumers, might help put dinner on more tables. He was right, and SPAM was soon being served to families across the United States. However, while a good ad campaign and a bad economy made SPAM a success in the United States, war would make it an international phenomenon.

World War II carried SPAM overseas. It fed American soldiers, but it would feed a lot more than the U.S. military, with the flow of SPAM saving people across Europe and the Asian Pacific from starvation. For much of the world, Uncle Sam became "Uncle Spam." In England, SPAM was welcomed and became nearly as pervasive as suggested by the famous Monty Python *Spam* sketch. Nikita Khrushchev noted in his biography that, in Russia, "without Spam, we wouldn't have been able to feed our army." By the end of the war, everyone knew about SPAM and though many may have complained during the conflict about the ubiquity of SPAM on the battlefront or in wartime rations, the world had developed a taste for the canned meat. Because it was wholesome, affordable, versatile, and easy, sales actually increased after the war.

The military presence in Hawaii led to Hawaiians becoming addicted to SPAM, and the state remains the largest consumer of SPAM in the United

Photo 8.1. The SPAMVILLE sign at this U.S. Army camp is a light-hearted response to the ubiquity of SPAM during World War II. Photo courtesy of Hormel Foods.

States, devouring more than seven million cans a year. In Guam, another island long associated with the U.S. military, annual consumption of SPAM averages more than twenty-four cans per person.

The involvement of the United State in the Korean War introduced SPAM there as well. Today, both in South Korea and in many Korean restaurants in the United States, one can find a dish called *budae jjigae*, which, roughly translated, means "Army Base Stew." It includes the familiar Korean ingredients of kimchee, noodles, tofu, red pepper paste, and green onions. It may also include sliced hot dogs, Polish sausage, baked beans, and American cheese, but it will always include thinly sliced SPAM. (This tastes much better than one might imagine, and for Koreans of a certain age, this is classic comfort food.) After the United States, Korea is the second largest consumer of SPAM.

While there are numerous stories about what the name means and where it came from, the official story is that Kenneth Daigneau, the brother of a company executive, blurted the word out at a New Year's Eve party, during a discussion of what to call the new product. In a 1945 interview for *The New Yorker*, Jay Hormel said he knew the moment Daigneau spoke it, that it was the right name for the product. As for what is in the can, originally it was all pork shoulder. Now, it is pork shoulder and ham. However, SPAM has also evolved with the times, reflecting not only changing tastes and trends but also the international popularity of the product. In addition to SPAM® Classic, one can now find (among others) teriyaki, chorizo, jalapeño, hickory smoked, black pepper, bacon, Portuguese sausage, or garlic SPAM, as well as SPAM®LITE (less salt and fat).

The world views SPAM as purely American, and, like other American imports, they are generally glad to have it. SPAM is now sold in more than forty countries. But there is still a constant demand in the United States. Sales rise each time the economy dips. Changing dietary concerns, demographics, and culinary trends may have led to new varieties, but SPAM continues to do what it has always done—what pork has always done—make protein cheap, tasty, and readily available.[16]

CHAPTER NINE

Local Pig

Influences and Specialties in the Heartland

Corn is one thread that holds together the fabric of American cuisines, and, of course, there is pork. Almost every people group that came to the United States (with obvious exceptions) contributed some pork application. Traditions were carried westward with settlers, and so popular dishes in the Midwest can often be traced back to the East, the South, or even to Europe. But regional differences began to develop, and dishes began to be associated with places.

By the end of the 1800s, the Midwest was the most ethnically diverse region in the country.[1] This chapter won't attempt to relate the full range of pork-based contributions to our culinary wealth, but it will look at many of those who came and what they contributed. It will also look at some quirky local specialties, from foods that wandered into the Midwest early in the region's history to those that were created or transformed here, whatever their parentage. (Recipes for a few specialties are in the chapter that follows.)

Looking through some of the many cookbooks that emerged during the Civil War and shortly thereafter, one generally finds the recipes hard to follow (measures and recipe writing were not standardized until the end of the 1800s). However, many of the dishes still popular across the Midwest and even across the country are easily recognizable. People have always wanted to eat well, so even in the mid-1800s and close to the frontier, cookbooks were dotted with curries, soufflés, oysters, pastries, ginger, lemons, and other taste treats, all reflecting the backgrounds of communities, as well as of the increasing access to exotic ingredients afforded by improved transportation.

Not too surprisingly, there were multiple recipes for pork, though not just for mealtime. In cookbooks from the mid-1800s, and even up into the early 1900s, instructions were included for such activities as curing hams, "trying" lard, smoking meat, brining, and making salt pork. Most cookbooks included directions for making headcheese, souse, pigs' feet, and other preparations that reflected careful stewardship of resources. Of course, there were also plenty of recipes in which familiar forms of pork were the stars—whole roasted pigs, baked hams, pork pies, grilled sausages, tenderloins and chops, bacon, and ribs. But even when some form of pork was not the focal point of a dish, it often appeared as an ingredient. "Add a slice of ham," "cook with bacon," and "boil with a pound of salt pork" were common instructions. One recipe for leftover turkey advises, "a little cooked ham minced fine and mixed with it is an improvement," and a fricasseed chicken is flavored with "a few slices of salt pork." Recipes that included salt pork appear on almost every page of these old cookbooks. Soups, especially bean soups, were flavored with it. Fish dishes were thought to be improved by it. Even a recipe for corned beef calls for a piece of salt pork to be added to the pot. (Salt pork was the main source of salt in cooking.) Include lard in the discussion, and everything with a pastry crust or that was fried can be added to the list of pig-reliant dishes. So, while cooking in the mid-1800s was generally varied and interesting, it was also tremendously dependent on pigs.[2]

Those Who Came

In the 1780s, as migration began into the region that would become known as the Midwest, it was almost entirely from the eastern states.[3] The first people to migrate in large numbers were from the Upland South, an area that stretched from southeastern Pennsylvania, across Virginia and the Carolinas, to northern Georgia.[4] These Southerners settled in Ohio, Indiana, southern Illinois, and Missouri. They brought their farming systems, along with their dependence on corn and hogs. Not all the first settlers were from the Upland South, but the majority were. However, since payment for service during the Revolutionary War was often the promise of land, there were also settlers from Pennsylvania, New York, and the New England states right from the start.[5] French Canadians were also part of the mix, since they had been in the region for more than a century as trappers and traders.[6]

Starting in the 1820s, folks from the Middle Atlantic region began moving into the Midwest. Eastern cities were crowded and expensive, and the promise of the open plains and cheap, fertile land was appealing. As transportation options improved, from the creation of the National Road to the building of canals and, finally, the rapid growth of rail systems, new waves of settlers migrated west from New England and the Middle Atlantic states. After two hundred years of living in widely separate and very different regions, there was something of a clash of cultures as New Englanders moved into an area primarily inhabited by Southerners.[7]

By the mid-1800s, news of the seemingly limitless farmland had circled the globe, and people began to arrive in the millions from Europe and Asia—most bringing strong traditions of reliance on pork with them. The advent of the potato blight, though most famously and devastatingly associated with Ireland, had an impact on several European countries, and potato-reliant Poles and Germans joined the Irish who were flooding westward in the mid-1800s. (It's worth noting that Germany as we now know it didn't exist at the time, so the term "German," when applied to groups from before the mid-1900s, actually refers to German-speaking people from Hanover, Pomerania, Saxony, Prussia, Austria, and more than two dozen other distinct provinces, principalities, or countries in northern Europe.)

While most people wanted farms, many got jobs in the region's rapidly growing towns. But whether in urban areas or rural, those coming from overseas usually settled near people who shared their language, culture, and foodways. Roughly half of all the Germans who arrived in the mid-1800s settled in a region anchored by Milwaukee, St. Louis, and Cincinnati.[8] Swedes, who had begun arriving in 1825, swept across the entire upper Midwest. Finns, too, fancied the upper Midwest, with Michigan becoming the heartland of Finnish America. Norwegians settled primarily in Minnesota, Wisconsin, and North Dakota. Danes chose farming regions of Minnesota, Wisconsin, Iowa, Illinois, and Kansas. German and Irish immigrants were still more numerous than Scandinavians, but Scandinavians arrived in large enough numbers to alter the culture of the American Midwest.[9] Enclaves of Dutch and Belgian settlers, who also arrived in the mid-1800s, were scattered around the region.[10]

The Homestead Act, passed in 1862, was designed to draw new settlers to the Great Plains, and it did. However, most Europeans waited for the end of the Civil War before considering homesteading. Then, once the conflict was over, the rush began again in earnest. A new wave of Polish immigrants headed for the Great Plains beginning in the 1870s, with major settlements in central Nebraska and eastern North Dakota, with smaller farming communities in Kansas and South Dakota.[11]

The Chinese came to the United States during the Gold Rush in the 1840s and then in even larger numbers to work on the Transcontinental Railroad in the 1860s. When attitudes toward the Chinese turned ugly on the West

Coast in the 1870s, the Chinese migrated to the Midwest, which was more welcoming. Chicago and St. Louis were the first Midwestern cities with Chinese populations. In time, Illinois, Missouri, Michigan, and Ohio all had significant Chinese populations.[12] By 1890, roughly six hundred Chinese lived in Chicago, about a quarter of them settling in the city's first Chinatown.[13]

Following emancipation, many African Americans took advantage of the Homestead Act to create all-African American farming communities. A dozen "colonies" were built in Kansas.[14] However, most moved to cities, and most came much later. The Great Migration saw the movement of more than six million African Americans from the rural South to the cities of the North, Midwest, and West from 1916 to 1970.

Between 1880 and 1920, the relentless torrent of European immigration came increasingly from southern and eastern Europe. Italians, Greeks, Czechs, Slovaks, Hungarians, Lithuanians, Romanians, Serbians, Croatians, and more made their way to the Midwest, though many of them were drawn to the growing cities of the region.[15] Italians escaping poverty in Italy in the 1880s settled mostly in Nebraska and Kansas.[16] Czechs were among the largest European groups to make their homes on the Great Plains. However, those who didn't have enough money even for the relatively low costs of homesteading often found work in the cities.[17] Others moved to the region's bourgeoning mining and lumbering areas.

A few Mexicans arrived in the region before 1900, but large Latino communities did not begin to form until the early 1900s. Around the turn of the century, both Michigan and Minnesota saw large influxes of Mexicans who arrived to work in the sugar-beet fields, though it was the cities that had the greater growth in Hispanic populations.[18]

One of the things all these groups had in common was foodways that utilized pork to at least some degree, and for some, it was the primary source of protein. Hence, the wealth of culinary options in the Midwest grew tremendously, and the region became as known for its diversity as for its abundance.

Specialty of the House

The English and northern European influences that had shaped dining in the East were carried westward by migrating settlers. The British contributed bacon, salt pork, hams, Sunday roasts, and the idea that a "proper" dinner included meat, a starch, and a vegetable. When there were regional specialties in the rural South and more urban North, these too would be carried westward and integrated into the developing Midwestern foodways. Among the many foods that moved across the country, barbecue and sausage gravy arrived from the South, while Yankee stews and baked beans were among New England's contributions.

Germans had been the largest non-English-speaking group in the colonies since the 1600s, so by the time the new wave of German immigrants reached the Midwest in the mid-1800s, German food was already so common as to not be thought of as ethnic food, or even as German food. *Delicatessen*, *pretzel*, and *hamburger* entered American English, and sausages and breaded, fried meats were pretty well integrated into the culture. (Just to underscore the influence of German-speaking Americans in the Midwest on the foodways of the entire country, consider that Irma Rombauer, who created *The Joy of Cooking*, was the daughter of Germans who arrived in St. Louis in 1860.)

While German influence is so well integrated as to be almost invisible in the Midwest, there are still plenty of sausages with names that make their origins clear: bratwurst, thuringer, braunschweiger, knackwurst, wienerwurst. Wisconsin in particular is known for producing an abundance of German sausages, though bratwurst is the best known—and Usinger's, opened in 1880, is still the top bratwurst maker in the state. Wisconsin not only makes but also consumes more bratwurst than any other state, often cooking them in the German-style beer for which Wisconsin is also known. And the state capital, Madison, hosts a four-day fundraiser that lays claim to being the world's largest Brat Fest.

(Polish sausages and Italian sausages are also big in the Midwest, though are generally more common in the bigger cities.[19] Not too surprisingly, one finds Chinese, Mexican, Hungarian, French, Russian, Czech, and everyone else's sausage, as well.)

Sausages were far from being the only foods Germans introduced—consider potato salad, sauerkraut, and strudel—and are even far from being the only pork-based foods. But even foods that were not originally pork-based could be transformed. Since pork was so abundant, German veal dishes (including bratwurst) were often converted to pork in the Midwest. In addition to food, however, Germans also introduced continental manners to the then still fairly wild West.[20]

Sometimes, dishes became associated with the places people settled or found work. In Minnesota's Iron Range, a region where many found work in the iron mines, recipes still reflect the ethnicity of those who came. *Porketta*, a pork roast seasoned with garlic and fennel and generally made from pork shoulder, is popular in the Iron Range, introduced by the Italians who settled there. (However, Iron Range porketta should not to be confused with *porchetta*, a more recently introduced Italian preparation. See Versatile Pig.) Also well known in the Iron Range is *sarma*. This dish of cabbage leaves stuffed with ground pork, ham, and beef, is traditional among Croatians in the area.[21]

Other groups similarly nurture cherished traditions. In Holland, Michigan, one can sample Dutch specialties such as *saucijzenbroodjes* (aka pigs in blankets), a pork sausage wrapped in flaky pastry, or thick pea soup loaded with pork. Indulge in pork *carnitas* or *cochinita pibil* in one of the myriad Mexican restaurants in Chicago. Choose between roast pork, baked ham, fried pork chops, smoked kielbasa, and boiled ribs at the Polish Village Café in Hamtramck, Michigan. Or one might opt for *char siu* (Chinese barbecued pork) in almost any of the major cities in the Midwest.

In Detroit, Chicago, St. Louis, and Cleveland, one can also seek out *soul food*, the label attached in the 1960s to foods brought north by migrating African Americans. A traditionally pork-centric cuisine, the African American traditions that were introduced from the South included pork chops but tended to focus on the cheaper items, such as ham hocks, pigs' feet, spare ribs, and chitterlings, as well as beans and greens cooked with pork. These foods were not simply a way to establish identity but also to find comfort, as the transition from rural life to urban was often rocky. While the term "soul food" did not exist until the 1960s, the foods that brought this comfort had begun appearing in Chicago as early as 1910.[22] And, while migrating whites from the Upland South had taken barbecue to Kansas City, it was African Americans who brought Southern barbecue to the urban North. In addition to places labeled "soul food," by the 1950s, nearly every city in the country had its share of black-owned barbecue restaurants.[23]

In addition to foods associated with different cultures or ethnic groups, there are also numerous foods that are tied to places across the Heartland.

A surprising number of sandwiches are clearly associated with specific cities. Springfield, Illinois, is home to the horseshoe sandwich, an open-faced sandwich that gets its name from the fact that when slices of ham are cut off the bone, the resulting shape looks like a horseshoe. To make the sandwich, put two slices of good ham on top of two pieces of toast, top that with French fries, and pour cheese sauce over the whole thing.

Visitors to Cleveland may encounter the Polish boy, a grilled kielbasa on a roll, with coleslaw, French fries, and hot sauce on top. St. Louis offers a crispy pig snoot sandwich. The meat in this sandwich is cut from the nose and cheek of the pig and is grilled until golden brown and crispy. Pig ear sandwiches are also popular in St. Louis, but they appear elsewhere in the Midwest. The Gerber sandwich is another St. Louis specialty. It's an open-faced ham-and-cheese sandwich made with garlic bread. The thing that limits it to St. Louis is that the cheese used, Provel, is sold only in St. Louis.

The cudighi sandwich is a ground-sausage sandwich from the eastern half of Michigan's Upper Peninsula. Cudighi is a spicy, Italian-style sausage, and the sandwich is served on Italian bread and topped with mozzarella and tomato sauce. Evansville, Indiana, is home to a sandwich that reflects the German culture of the region, along with the traditional "use it all" attitude: the pig brain sandwich. It's a pig brain that is battered and fried whole and served on a bun. A city-specific obsession that is not a sandwich is Cincinnati's goetta, a breakfast sausage made with pork and oatmeal.

The fried pork tenderloin sandwich is not associated with a city, but rather with two states (and trailing into surrounding areas). Both Indiana and Iowa consider this to be a defining dish.

In Ohio, a popular local specialty is sauerkraut balls. These are meatball-sized fritters that contain sauerkraut and some combination of ham, bacon, and pork. Rather than being a tradition from the Old World, these are more a blending of ideas from a variety of ethnic cultures, and while the recipe was invented in the late 1950s by Max and Roman Gruber for a restaurant in Shaker Heights, Ohio, these morsels have since migrated across the state.

In Iowa, in addition to the tenderloin sandwich, there is an Iowa Chop. Like the Ohio sauerkraut balls, this is a recent creation. Because Iowa is the top pork-producing state in the country, the Iowa Pork Producers and Iowa Development Commission came to the conclusion that the state should have an exceptional cut of pork to call its own. The cut they came up with was the Iowa Chop, which has to be a fresh twelve- to fourteen-ounce bone-in center loin chop that is 1.25 to 1.5 inches thick. The name and specifications were registered in 1976 and trademarked in 1982, to make certain that it would remain a symbol for the state. At the Iowa State Fair, more than ten thousand Iowa Chops are regularly consumed during the eleven days of the event (along with a full range of other pork specialties).[24]

Today, while some of these items may be unfamiliar outside their original locations, many of them are well known across the region and even beyond. It is unlikely that most people even think of ethnicity when they approach grilled bratwurst or barbecued spareribs. Still, the regional specialties contribute to creating community, and we can hope that knowing where the foods came from will help us remember the histories of those who brought them.

CHAPTER TEN

Transformed Pig

Recipes for Specialties

The people who came to the Midwest contributed both specific dishes and general influences to the developing cuisines of the growing nation. Even in communities with shared heritage, recipes evolved in the new land, where familiar ingredients might be hard to find. Some dishes remained associated with specific ethnic groups, but there were also foods that, thanks to millennia of trade routes and invasions in the Old World, were already common in a wide range of countries. Still others arose in various parts of the United States. Foods began to be shared at church socials, political rallies, and through the cookbooks that became increasingly common in the mid-1800s. In time, it became simply American food.

The recipes that follow are just a small fraction of the possibilities for the region. They are everyday fare, the things that people ate (and still eat) when they wanted to enjoy a meal, rather than impress a guest. Some have traveled, some are closely associated with specific places. All are delicious and, while anchored in history, all are still popular and being made in homes and restaurants around the Midwest.

Sausage Gravy

Sausage gravy is a classic Southern breakfast item, traditionally served over freshly baked biscuits. That said, because so much of the Midwest was settled by folks from the Upland South, it's common fare in the southern half of the Heartland and can even be found at diners in Chicago. It's cheap and flavorful, and it was a wonderful way to feed a hungry family, especially on a farm, where sausage and milk would almost always be readily available. Easy to make in large quantities, it was also a popular way to feed workers in the lumber trade, which is why it is sometimes called sawmill gravy or camp gravy.

I have always loved sausage gravy, even before I understood its significance or history. There are many regional variations, but the version I grew up enjoying was the white, milk-based gravy. Some people add garlic powder or cayenne, to zip it up or cut the richness, but since the black pepper and sausage are the flavors that delight me in this dish, I opted for one of the earlier and simpler versions.

1 pound bulk pork sausage
⅓ cup all-purpose flour
3½ cups whole milk, or more as desired
½ tsp. salt, or to taste
1 or 2 tsp. freshly ground black pepper, or to taste
Biscuits—your favorite recipe or packaged

Crumble the sausage into a large skillet. Brown the sausage over medium-high heat, continuing to break up the lumps. Cook, stirring, until no pink is visible. Reduce the heat to medium-low. Sprinkle half the flour over the sausage, and stir until flour is no longer visible. Then add the second half of the flour, and stir to thoroughly combine. (Do *not* drain the fat. The fat plus flour creates a roux that is the base of the gravy.) Pour in the milk, and stir to combine, being sure to scrape the bottom of the pan for any bits of sausage that might have stuck.

Cook the gravy, stirring frequently, until it becomes thick and luxurious. This can take ten minutes or more, but watch carefully, as it can happen fairly suddenly. Add more milk if it gets too thick too quickly, or if you want to stretch this to serve more people. (I've added an additional ½ cup of milk with no adverse effect, and I suspect one could add a bit more.) Taste the gravy. Stir in the salt, if needed (sausage may be salty enough), and pepper. (**Note:** 2 tsp. pepper, while perfect to my taste, might be more than some enjoy, especially if really good, fresh pepper is used, so you may want to add 1 tsp. pepper, then taste before adding more.)

Warm the biscuits, split, and ladle on the warm gravy.

Serves six to eight.

City Chicken

Long before an ad campaign touted pork as "the other white meat," it was being used as a substitute for chicken or veal. Chickens, which originated in the tropics, regularly failed to survive winter, so in northern states, chicken, as well as eggs, were costly luxuries. It wasn't until the 1920s that it was discovered that vitamin D would keep chickens alive during the winter, when sunlight was reduced. So, a politician in 1928 promising a chicken in every pot was a big deal. Before that discovery, people turned to the animal they had always eaten when price was a consideration: pork.

The pork is cut up and skewered, to make it look something like a drumstick, and city chicken is often called mock drumsticks. Eastern Europeans were responsible for the evolution of the recipe, and the region where the dish became important was in urban areas in the so-called Rust Belt, which stretched from New Jersey and New York across Pennsylvania and Ohio, to industrial centers in Michigan, Indiana, Illinois, and Missouri. As with any dish born of necessity, rather than of a chef's inspiration, there is no single "authentic" recipe; there are dozens, if not hundreds. All reflect personal taste, cultural background, and/or what was on hand at the moment.

In keeping with the Eastern European influence, this might be served with a sour cream sauce of some sort, mustard, or gravy, and possibly a side of *haluski*, a dish of fried cabbage with noodles, or perhaps dumplings. Today, in the eastern part of the Rust Belt, some stores still sell packages labeled "city chicken," with both cut-up pork and skewers enclosed. Pork shoulder is the most traditional cut, and it works well in this dish, because it's a bit tough and has enough fat. Don't use anything too lean or too tender.

Happily, if you don't need four servings, this reheats beautifully and freezes well.

1½ lb. boneless pork roast
½ tsp. salt, or to taste
¼ tsp. ground black pepper, or to taste
½ tsp. dried thyme
8 6-inch skewers
⅓ cup all-purpose flour
2 eggs
2 tbs. water
1 to 1¼ cups breadcrumbs (see Note 1)
2 cups vegetable oil for frying

If using bamboo skewers, soak them in water for half an hour, so they won't burn. (And if you can't find six-inch bamboo skewers, pruning shears can quickly reduce longer skewers to the right length. Metal skewers are fine, as well, but will cook meat a little faster, as they conduct heat.)

Trim pork and cut into pieces that are roughly 1 to 1½ inches cubed. Sprinkle pork with salt, pepper, and thyme. Thread pork onto skewers, pushing pieces firmly together. For the final product to look a bit more like a drumstick, place smaller pieces at the bottom of the stack.

Place the coating elements in three shallow dishes (pie plates work well). Place flour in first dish. Whisk eggs with water in second dish. Place breadcrumbs in third dish. Roll the skewers, one at a time, in the flour, coating well, then shake off excess. Dip the skewer in the egg mixture, and then roll each skewer in the breadcrumbs. Press the breadcrumbs into the meat, to make as much as possible stick, at the same time shaping the meat so that it looks roughly like a single piece of meat. Place coated skewers in the fridge for fifteen minutes, to firm up.

Place a wire rack in a baking dish or rimmed baking sheet. Position oven rack in center of oven and preheat oven to 325°F. Heat oil in a large skillet until 350°F. (If you don't have a thermometer for checking the heat, see Note 2 below.) When oil is hot, fry the skewers in batches (to avoid crowding) until golden brown, about two minutes on each side. Place the fried skewers on the rack and place baking sheet in oven. Allow to bake for thirty minutes, remove from oven, and turn skewers over. Return to oven and bake for another twenty minutes, or until all meat is tender and none is pink in the center. (The additional twenty minutes may be unnecessary if you used lean or tender meat, so you might want to check a piece after the first thirty minutes.) When done, allow skewers to rest for five minutes before eating. Good plain, but can be enhanced by a mustard sauce or sour cream sauce. Serves four.

Note 1: Traditionally, the breadcrumbs would be homemade from day-old bread. Closest to this would be regular packaged dry breadcrumbs. However, a lot of modernized versions use Panko breadcrumbs, to make this crunchier and a bit more upscale. What this means is you can use whatever you have or prefer. It all works.

Note 2: If you don't have a thermometer that goes up to 350°F (frying thermometer or candy thermometer—don't use a meat thermometer), just be sure to put the oil in a cold pan and then turn the heat on medium. Watch the oil. In a few minutes, it will begin to shimmer. Depending on the oil, it may begin to smoke ever so slightly. Soy oil and canola oil smoke around 350°F, vegetable shortening at 360°F, and lard at 370°F. It actually only takes a few minutes, so don't walk away after you turn on the heat. Once the oil starts shimmering, you can begin testing it. Drop in a pinch of breadcrumbs or flour, and if it sizzles and acts like it's being fried, you're ready. (Peanut oil has a smoke point of around 440°F—so if you want to cook with peanut oil, buy a thermometer.)

Iron Range Porketta

The Iron Ranges are a collection of massive iron-ore deposits situated around northern Lake Superior. The ranges actually stretch across Michigan's Upper Peninsula, northern Wisconsin, and into Ontario—but when most people speak of the Iron Ranges, they're thinking of Minnesota. It was in Minnesota that iron mining became an industry of epic proportions—not just employing thousands, but also supplying the iron for the steel mills that built the nation's cities and helped the United States win two World Wars.

Immigrants began flowing into Minnesota in the late 1800s, to work in the mines. As more ore was found, more mines started, and more people came. Nearly two dozen nations were represented in the region's rapidly expanding population, from Eastern Europe, Southern Europe, Western Europe, Scandinavia, and the British Isles. Several regional specialties in northern Minnesota reflect these varied backgrounds.[1]

Iron Range porketta arose from the region's Italian population, but today it is nearly ubiquitous throughout Minnesota's Iron Range. The roast cooks for a long time at low heat, which leads to meat that melts in mouth. This dish requires some advanced planning, as it needs to be refrigerated overnight.

2 tbs. fennel seeds
2 tsp. salt
2 tsp. ground black pepper
2 tsp. granulated garlic
handful Italian flat-leaf parsley, finely chopped
4-lb. boneless pork shoulder (aka Boston butt) roast
1 fennel bulb
salt and pepper to taste

To serve:
8 Italian hard rolls (or other crusty bread rolls for sandwiches)
mustard

Toast the fennel seeds in a frying pan for a couple of minutes, until they become fragrant. Then grind them with a mortar and pestle or in a spice grinder. Combine the ground fennel with the salt, pepper, and garlic.

Slice through the roast parallel to the cutting board, but stopping about half an inch from the edge. Then open the roast up and flatten it—that is, butterfly the roast. Do *not* trim the fat. Then cut quarter-inch deep slits in the meat every inch or two, changing direction to create a diamond pattern. Sprinkle the spice mixture over this open side of the roast, rubbing it in so that the spices get into the slits. Then sprinkle the chopped parsley over the opened roast.

Roll the roast up tightly and wrap in plastic wrap. (If the roast doesn't cooperate, you can tie it closed, but that may be unnecessary.) Refrigerate overnight (or for a minimum of six hours).

When ready to cook the roast, unwrap it and place it fat side down in a roasting pan. Preheat oven to 325°F. Split the fennel bulb, remove the stalks and discard (though you may want to save the fronds, as they are very flavorful), split the bulb in half, remove the core, and then chop roughly. Spread the chopped fennel evenly over the top of the roast. Cover the roasting pan tightly with aluminum foil. Roast the pork for three hours or until the meat registers 200°F on a meat thermometer and a fork slips easily in and out of the meat.

Move to a platter or cutting board and allow to rest for thirty to forty-five minutes. Taste for seasoning and adjust if needed. Shred and serve on the hard rolls—traditionally with mustard. Serves eight.

Optional step: Strain the liquid in the roasting pan. Use a fat separator to eliminate excess fat. Toss ⅓ cup of the defatted cooking liquid with the shredded pork before serving. Otherwise, just chill the liquid, peel off the fat layer, and use the porky, garlicky gelatin to flavor vegetables or rice.

Split Pea Soup

Pea soup arrived with almost everyone from Northern Europe, as well as with some from Eastern Europe. Yellow split peas are almost as common as green split peas in these recipes, and the meat might be ham, smoked pork hocks, or sausage. Yellow peas (which, unlike green peas, have to be soaked in advance) appear in Polish and French pea soups. The Dutch use green peas, but they might add spareribs or ham or, for a heartier soup, pork chops and Dutch sausage. Made with green peas and salt pork, however, split pea soup is most closely associated with Britain and Ireland. That said, there are Scandinavian recipes that are similar to the one below.

Good, thick pea soups are comforting, and while in their various forms they are familiar to most in the Midwest, their history stretches back well into antiquity. Pea soup was popular enough as a street food in ancient Athens that it appeared in many of the works of the Greek playwright Aristophanes.[2] The soup, though delicious, was long associated with poverty, because the ingredients are so inexpensive.

Salt pork, which is essential to many traditional dishes, is available in most grocery stores. Look for good, meaty salt pork. ("Meaty" as in something like meaty bacon; it's made from pork belly, so it will never be mostly meat.) Don't worry about the fat; no more will get into the soup than if you added oil to sauté the onions. It will simply help flavor the soup. If the salt pork you find is crusted with salt, rinse it off before using it. However, most commercial, packaged salt pork won't have this problem.

Midwestern cookbooks from the 1800s generally include several recipes for split pea soup. The versions might not vary greatly, but every recipe contributed to these fundraising books got included, and almost everyone had a recipe for pea soup. Carrots and celery might get added in later recipes, as gardens flourished, but this version is something that could have been prepared with items that were carried along in the wagons that headed for the frontier. It also happens to be quite wonderful. So much better than the stuff that comes in cans.

½ lb. meaty salt pork
1 large onion, chopped
2 cloves garlic, minced
1 lb. green split peas, picked over and rinsed
8 cups water
1 bay leaf
1 tsp. salt
1 tsp. dried summer savory or marjoram
½ tsp. black pepper, or to taste

Cut the salt pork into one-inch cubes. In a five-quart stockpot, fry the salt pork over medium-low heat until the pork has begun to brown and has given up a good bit of fat (at least a couple of tablespoons' worth). Add the onions and garlic, and cook gently for ten minutes, or until onions begin to take on a little color.

Add the peas, bay leaf, and eight cups of water to the pot. Bring to a boil. Skim any foam that rises to the surface. Reduce heat and simmer for one hour. Add salt, herbs, and black pepper, and simmer for an additional thirty minutes. Toward the end of the cooking time, stir frequently, to avoid scorching the soup. If the simmer has been too vigorous, and the soup appears to be turning into a solid, add a bit more water.

Remove the bay leaf and cubes of salt pork. Separate the meat from the fat of the salt pork, shred the meat, and add it back to the soup. (There won't be a lot of meat, but it's still enough to enhance the soup. The fat can either be rendered for later use or disposed of.) Taste the soup for seasoning and adjust if necessary. Serves six.

Note: Commercial salt pork usually comes in twelve-ounce packages. Since this recipe calls for a half pound, you'll have four ounces left over. If you don't need to save it for another recipe, then you can just dice it and fry it over medium heat until crisp, and use it as a garnish for the finished soup.

Goetta

Goetta (pronounced "get-uh") is a type of breakfast sausage that has a devoted following in Cincinnati, Ohio, and the surrounding region. It is sufficiently popular that the annual, four-day Goettafest regularly draws more than one hundred thousand people, who come to play games, listen to music, and dine on goetta-based dishes, such as goetta gumbo, goetta sushi, goetta nachos, goetta fried rice, and the simple overindulgence of deep fried goetta balls.

Goetta first appeared in the Cincinnati area in the early 1800s, at a time when the city was known as "Porkopolis." The combination of a large German population and Cincinnati's tremendous pork-processing industry may have made its creation inevitable.

Goetta is a member of the large family of economical, German-influenced foods that includes both scrapple and gritswurst (also known as gritz)—dishes that, like goetta, use a grain to make the most of a small amount of meat. These dishes start with pork, but the grains vary: the grain in goetta is oats; in scrapple, it's cornmeal; and in gritswurst, it's usually buckwheat, though gritswurst sometimes employs a barley-and-oats blend, or even just oats, with region, rather than grain, determining the name of the dish. Goetta is technically a sausage, but it might be easier to imagine if you think of it as a good meat hash, warm, savory, with crispy edges. Perfect on its own, but also a great perch for a nice poached egg.

While modern recipes often blend beef with pork, the more traditional version—the version that makes sense for people trying to eat cheaply and living in a town known for pork—uses pork alone. That said, the pork used can range from trimmings to grinding up pork shoulder to simply buying a package of ground pork. Some versions have the raw meat cooked first, to make a broth, grinding the meat after it's cooked, and cooking the oatmeal in the broth before combining the two. Most versions today, however, start with ground meat, adding it to seasoned, cooked oatmeal, as in the recipe below.

Recipes for goetta will list pinhead oatmeal among the ingredients. Pinhead is actually another name for steel-cut oats—oats that are cut rather than rolled. Pinhead/steel-cut oats are much heartier than rolled oats. They take longer to cook, but in this recipe, that's vital. *Make sure you do not buy rolled oats*—and also make sure the steel-cut oats you buy are not "quick cooking oats." You want regular, old-fashioned, slow-cooking, steel-cut oats. As for the pork, don't get anything too lean, as the fat in the pork is what makes it possible to fry this without added oil.

This dish is so popular on its home turf that you'll generally find recipes that make anywhere from six to ten pounds of goetta. The recipe below gives you one loaf-pan worth, so you can try it out and decide if you like it. It's actually kind of addicting, so you may be glad to learn that this recipe doubles easily—and it is a great way to make meat go further.

Of course, if you live in Cincinnati or any of the nearby parts of Ohio, Indiana, or Kentucky, you probably already have a recipe for this—or else you're buying one of the several commercial brands available at most grocery stores. In that case, I don't need to convince you that this is worth eating.

4 cups water
1¼ cups steel-cut oats
1½ tsp. salt
⅛ tsp. black pepper
1 lb. ground pork
1 medium onion, chopped
1 clove garlic, minced
1 to 2 bay leaves

Pour the water into a three-quart saucepan and bring to a boil. When boiling, stir in salt, pepper, and oatmeal. Lower the heat to a simmer, cover, and cook for forty-five minutes, stirring the oatmeal regularly.

Add the pork, onion, and garlic, and stir vigorously to break up meat, so that all ingredients are well blended. Add the bay leaf. Cook, covered, over low heat, for 1 to 1½ hours. Stir about every fifteen minutes initially, but increase frequency toward the end of the cooking time, as mixture thickens. Goetta is fully cooked when there is no longer any liquid pooling on top of the mixture. Remove bay leaf (or leaves) and pour mixture into a loaf pan. Cool and place in refrigerator for at least two hours, though overnight is better.

When ready to eat, slice goetta fairly thinly (about ⅓- to ½-inch thick), and fry over medium heat until well browned and crispy. If meat is fairly lean (as is common with pre-packed ground pork), add a bit of oil or bacon drippings to the pan.

Serve goetta on its own, with eggs any style on the side, or with something sweet, such as applesauce or maple syrup. Depending on how thickly you slice it, if you have two slices for breakfast, it serves six to eight. It will keep in the fridge for up to six days, and can be frozen for up to a month.

Smalec

Among the traditions Eastern Europeans brought with them to the United States was reliance on "poor man's butter," a spread made with lard. Lard spreads were and are enjoyed by Czechs, Poles, Slovaks, Hungarians, Lithuanians, Croats, Serbs, Slovenians, Bulgarians, and Romanians. (Though cherished delicacies made from lard can be found throughout Europe.)

While people came to the United States from all these countries, it was the Poles who came in the greatest numbers. Substantial Polish communities settled in New York and Chicago, and continue to grow—with New York only recently surpassing Chicago as having the largest Polish population outside of Poland. Wisconsin, on the other hand, has more communities of Poles, rather than a large urban population. However, Poles have settled in lesser concentrations across the country. As a result, *smalec*, Poland's version of Europe's nearly ubiquitous lard spread, is a fairly commonly encountered version of this traditional treat. When bread is served in Polish restaurants, it is generally accompanied with *smalec*, and Polish markets will more than likely have abundant *smalec* in their deli sections.

Recipes for *smalec* are numerous, varied, and often make large quantities. The version that follows reflects the most common variation, and while it does not make a huge quantity, it makes enough that it's good to know it freezes well. If you don't have a meat grinder, ask your butcher, and if you don't have a butcher (or the butcher can't do it), know that while dicing the fat can be tedious, your results will still taste great. But note that finely diced fat will produce a slightly chunkier *smalec*, and it will take a little longer for the fat to render out. It will still be delicious. And while the marjoram is traditional and very tasty, it can be left out, if you don't have any available.

The full name of the dish in Polish is *Smalec ze Skwarkami*, meaning Lard Spread with Cracklings. While this is definitely not a diet food, there is some comfort in the fact that lard is lower in cholesterol than butter. Not health food, but not as bad as one might think. The extra calories afforded by fat were once necessary for survival, but this spread continues in use for one simple reason: the taste is sensational.

1½ lb. white pork fat, ground or finely diced
½ lb. smoked, fatty bacon, finely diced
1 large or 2 medium onions (about a pound)
2 or 3 cloves garlic, minced
1 large, tart apple (Granny Smith is a good choice), grated or finely diced
1 tsp. salt, or to taste
1 tsp. dry marjoram or a few sprigs of fresh
⅛ tsp. finely ground black pepper

Place the pork fat in a large skillet over medium heat and fry until fat is transparent and rendering its fat (usually fifteen to twenty minutes, if fat has been ground). Add the bacon, onion, and garlic to the fat, and stir together. Continue to cook for roughly fifty minutes to an hour, stirring occasionally, until most of the fat is rendered and the bacon is crisp and brown. Add the apple, salt, pepper, and marjoram. Stir to combine. Cook for two to three additional minutes, until apples are softened. Remove from the heat.

When *smalec* is cool enough, taste and add more salt, if needed. The bacon may have added enough, but you want this to be flavorful. Put in a container (metal, glass, or crockery are best). Stir as it continues to cool, so that the cracklings are distributed throughout the spread, rather than sinking to the bottom. Serve with good rye bread. Will keep in the refrigerator for up to two weeks.

If you want to stretch this a bit, you can add another ¼ lb. of white pork fat without appreciably altering the final taste or consistency.

Collard Greens

Collard greens are members of the *Brassica* genus, which includes all the variations on the cabbage theme, as well as mustard plants. They were brought to the Americas from England, though these ancient vegetables had been widely disseminated, including to Africa, long before Europeans considered sailing west. As a result, enslaved Africans who were brought to the New World knew just what to do with the leafy greens that the British had planted here. Like so many other dishes that arose in the South during the years slavery existed, collard greens went from being the food of everyone to the food of the poor to a food primarily associated with African Americans.

It was during the Great Migration that collard greens began appearing in the big cities of the Midwest, especially Chicago and Detroit.

I love this dish. And I love drinking the "pot likker" (liquor) that it creates. Some recipes for this dish use chicken broth. However, this is not as traditional as just relying on the flavor of the ham hocks. Besides, the smoked pork adds such wonderful flavor, the broth is really unnecessary.

When you buy your collard greens, look for bright green leaves. Avoid yellow, pale, or spotty leaves. Collard greens are available throughout the year, but January through April are when they're at their best.

2 ham hocks (sometimes called smoked pork hocks), about 1½ lb.
1 tsp. granulated garlic, or to taste
1 or 2 bunches collard greens (about 1½ lb.)
1 medium onion, coarsely chopped
2 tsp. salt
pinch red pepper flakes, or to taste
¼ tsp. ground black pepper
1 tbs. drippings, lard, or other form of fat (butter okay, if drippings unavailable)
pinch of baking soda
1 tbs. apple cider vinegar
hot sauce to taste

Place the hocks and granulated garlic in a large pot with three quarts water. Bring to a boil. Reduce heat and simmer for one hour.

Wash the collard greens carefully, then fold each leaf in half lengthwise and use a sharp knife to cut off the stiff spine, which you can discard. (If you get very tender, young leaves, you don't need to remove the spine, but very tender leaves are more likely to be at a farmers' market than at a grocery store.) Once the spines are removed, stack several leaves, roll the bunch of leaves, and slice the roll at roughly half-inch intervals. This will give you long, thin strips of greens.

Place the greens, the chopped onion, the salt, red pepper flakes, black pepper, and fat into the water with the meat. Add the pinch of baking soda (this helps tenderize the collard greens). Simmer for forty-five to sixty minutes, stirring occasionally. (I set a timer for fifteen minutes, and the first two times, I just stir. Then, at the end of the third fifteen-minute period, I test the greens—because often they are tender enough by this point. If still tough, they can be cooked an additional fifteen minutes.)

When the greens are done, take out the ham hocks and set them aside. Stir the vinegar into the greens. When the hocks are cool enough, strip off the meat, breaking it into small pieces, and add it back to the greens. To serve, spoon the greens and pork out of the pot—but whatever you do, *do not drain them*. The delicious pot likker that is left behind is actually considered by many to be the best part of preparing this dish. It can simply be drunk or put in a bowl and sopped up with bread. But don't miss out on it. (Besides, after all the simmering, this is where most of the nutrients from the greens will be found.)

Traditionally served with hot sauce. As a side dish, serves six (not including the pot likker). With cornbread, it can make a meal, in which case it offers only three or four servings.

Breaded Pork Tenderloin

In many parts of the Midwest, if you ask what a typical local dish is, the answer will be breaded pork tenderloin, also known as BPT or just fried tenderloin. This thin, crispy, sprawling bit of meat is most commonly served as a pork tenderloin sandwich, which is most strongly associated with Indiana and Iowa, though it's popular across the Midwest and even beyond, and states across the region vie for the honor of saying they have the best version.

In South Dakota, they have pork fritters, which are essentially the same thing, but they're served on a plate with gravy, rather than in a sandwich. In Wisconsin, you'll find pork schnitzel, but that's a flattened pork chop, rather than tenderloin. That said, some people slice up pork loin, which is a row of pork chops with the bones removed, to make their "tenderloin" sandwiches—though flattened, breaded pork chops also appear as Milanese on some Italian menus. So, the lines blur. But technically, a breaded pork tenderloin is supposed to be made with tenderloin—a much slimmer, more tender piece of meat that sits under the loin on the pig. It should only be a couple of inches in diameter. If the piece of meat you're contemplating is four or five inches across, it's pork loin. It will still be good, just not quite as tender, and it may need to cook a bit longer than tenderloin.

While the influences behind frying pork may be old and varied, the title of inventor of the breaded pork tenderloin sandwich has been claimed by Nick's Kitchen in Huntington, Indiana. Introduced at the dawn of the 1900s, the recipe created by Nick Freienstein is still served at the restaurant, unchanged. That said, while the recipe might not have changed at Nick's, as with any popular dish with a bit of history, there are many variations. Some involve marinades or different seasoning combinations, some are baked or broiled, though most are fried, some are coated in crushed saltines, or just breadcrumbs without flour. But all include pork that has been pounded thin before being breaded.

In Iowa and Indiana, this is generally served on a Kaiser roll or good hamburger bun, preferably toasted, usually adorned with lettuce, tomato slices, onion, mayo, and mustard, with some folks adding ketchup. It might also be garnished with a few dill pickle slices. Of course, fried tenderloin can be served without the bun. It's still delicious—and it then belongs to an even wider range of food traditions.

1 pork tenderloin, about 1 to 1¼ lb.
1 egg
½ cup milk
1 cup all-purpose flour
½ cup dry breadcrumbs
1 tsp. salt
½ tsp. ground black pepper
½ tsp. onion powder
½ tsp. garlic powder
1 tsp. dried marjoram (optional; see Note)
vegetable oil for frying (probably about a cup)

Trim the tenderloin of excess fat and the silver skin. (Silver skin is a tough, whitish bit of connective tissue that lies along one end of an untrimmed tenderloin. Just slip a sharp knife under the silver skin and slide it along, as though you were shaving the tenderloin.)

Cut the tenderloin into four equal pieces. Butterfly each piece. (To do this, cut the meat as if you were cutting more pieces, but this time, only cut about three-quarters of the way through, then lay it open, like a book, and flatten it with your hand.) Put each piece of butterflied pork between two pieces of plastic wrap and, using a meat mallet or a fifteen-ounce can of something heavy, such as beans, gently pound/roll/stroke the meat until it is about ¼-inch thick.

To bread the tenderloin, set out two pans. Add the egg and milk to one pan, and whisk until well combined. Put the flour, breadcrumbs, salt, pepper, garlic powder, onion powder, and marjoram into the other pan, and stir to combine.

Dip each piece of tenderloin in the egg-and-milk mixture, turning it over so that it is well moistened. Then dip the meat in the seasoned flour mixture. Turn it over in the flour and press it down, to make sure the flour and crumbs stick.

Put oil in a large frying pan. Oil should be about a ½-inch deep, so the amount of oil will depend on the size of the pan. Heat the oil over medium heat until it reaches 365°F. (If you don't have a thermometer, see Note 12 that follows the City Chicken recipe.) Once the oil is hot enough for frying, put in one of the tenderloin pieces, making sure to lay it away from you, so you don't get splashed by the hot oil. If there is room in the pan, add a second piece. Don't put all pieces in at once, as it will lower the temperature of the oil.

Fry each piece for roughly three to four minutes per side, or until golden brown. Remove the fried tenderloin to a paper towel to drain. Create your sandwich or add side dishes, as desired (but note, if not making a sandwich, without the mustard and mayo, you might need to add a little salt to the fried tenderloin, or serve with gravy or sauce). Serves four.

Note: Marjoram is the classic herb for German pork dishes. Its German name is *wurstkraut*—the herb you use in wurst, or pork sausage. Marjoram is sweeter than oregano, but if you don't have marjoram on hand, you can substitute oregano. Or you can make this without any herbs, just the onion, garlic, and salt.

Ham Balls

Grinding meat has always been a popular way of making certain nothing was wasted, as you could add in all the scraps or oddly shaped pieces that didn't meet the needs of other recipes—and even bits of offal might be included, when some was on hand. We're not referring to sausage here, but to meatloaf, meatballs, burgers, croquettes, and other homey offerings that are generally as economical as they are tasty. The ground meat would be shaped and then baked or fried, producing a fine, meaty dish.

Ham balls appear to have a primarily Scandinavian heritage. Carried to the Midwest, ham balls (and other ground meat dishes) found their ways into early regional recipe collection. Here's a recipe from *Buckeye Cookery and Practical Housekeeping*, a collection dedicated "To the Plucky Housewives of 1876, Who Master Their Work Instead of Allowing It to Master Them."

The contributors of recipes were generally identified, and this one was attributed to Mrs. Howard Evans.

Ham Balls

Chop fine cold, cooked ham; add an egg for each person, and a little flour; beat together, make into balls, and fry brown in hot butter.[3]

Fortunately, because ham balls are still popular in the Midwest, there are lots of recipes that are not only easier to follow, but also updated for modern kitchens. Most kitchens now have ovens and an easy way to control temperature, and newer recipes all seem to take advantage of this. However, despite other differences, they all start, as the 1876 version did, with chopping up cold, cooked ham.

These ham balls are not toothpick fare—think mini ham loaves (full-sized ham loaves being another regional favorite). The balls are somewhere between golf balls and baseballs in size. These are popular all across the Midwest, but are particularly common in rural Minnesota and South Dakota, and across Iowa. Breadcrumbs are often replaced with crushed soda crackers in Minnesota and South Dakota. However, in Iowa, the tradition is to use graham-cracker crumbs in lieu of breadcrumbs, which creates a much sweeter ham ball. The sauce that is included with this recipe is similar to a nice, sharp, sweet-sour barbecue sauce. For the chopped ham, make sure it has some texture. You don't want a purée.

1 lb. finely chopped/diced cooked ham
½ pound ground pork
1 cup dry breadcrumbs
¼ tsp. black pepper, or to taste
¾ cup milk
2 eggs, well beaten
Sauce:
⅓ cup firmly packed dark-brown sugar
½ tsp. dry mustard
½ cup tomato catsup
1½ tsp. cider vinegar

Combine the ham, pork, breadcrumbs, pepper, milk, and eggs, mixing thoroughly. (Using your hands is probably the easiest way to do this.) Divide the meat into eight portions and roll each portion into a ball. Place the balls in a baking dish that will hold them comfortably, without squishing. (They don't need a lot of room around them, but you want them to bake evenly. I find that a dish of eight by eight by two inches works well.) Let them sit for a few minutes, so everything blends, and while they sit, preheat the oven to 350°F. When the oven is ready, place the ham balls in the center of the oven and bake for one hour.

While the ham balls bake, add the dry mustard to the brown sugar, and blend lightly (this keeps the mustard powder from forming lumps in the sauce). Then, whisk the catsup and vinegar into the brown sugar and mustard. When the hour is up, take the ham balls out of the oven and spoon the sauce over them. Then, turn the oven down to 300°F, return the ham balls to the oven, and bake for another ten minutes, or until the sauce is beginning to thicken. Serves four. (I like to serve this with rice or potatoes.)

The Midwest has far more to offer than just pork dishes—but also has vastly more pork dishes than those listed above. But I hope this sampling inspires you to try something you haven't tasted before.

SECTION III

LIVING WITH PIGS TODAY

CHAPTER ELEVEN

Popular Pig

The Rise of Pig Obsession

After the late twentieth-century fat phobia began to subside, pork found itself at the center of an unfettered counter-revolution. While many simply returned to fearlessly enjoying reasonable amounts of fat, the pendulum carried others into the realm of wild abandon. Pork belly found a place on menus at numerous upscale restaurants. Bacon was dipped in batter and deep fat fried. On television, food shows rhapsodized about pork. Anthony Bourdain's phrase "porky goodness" became common in food chat groups. From high-end dining to street food, the pig in its multitudinous culinary applications began appearing more frequently and was enjoyed more enthusiastically. At the same time, discussions of snout-to-tail eating ceased to be relegated to food histories, and people began trying long-ignored parts.

While any trend is complex, with multiple triggers, a rocky economy did help promote the resurgence of indulgence in delectable foods. As George Orwell noted in *The Road to Wigan Pier*, "A millionaire may enjoy breakfasting off orange juice and Ryvita biscuits; an unemployed man doesn't. . . . when you are underfed, harassed, bored, and miserable, you don't want to eat dull wholesome food. You want something a little bit 'tasty.'" Tasty is definitely what folks are seeking, and few things afford more tasty options than pork.

Of course, we don't just like pigs on the plate. Pigs pop up regularly in art, crafts, and decorative items. Pigs are comfortably frumpy. Unlike horses, dolphins, or eagles, pigs do not remind us of our limits and lack of grace. And the little ones are cute. But mostly, we like pigs because they taste good.

Just as porcine popularity has taken many forms, so have the outlets for showing appreciation for pork products. Long-standing events, such as state fairs, have seen increased attendance, new festivals have been started, competitions expanded, and old traditions have been resuscitated. People have gotten involved with pigs on many levels, even if it's just seeking new places to enjoy "porky goodness."

Cultural Pig

Of course, for many, pork never left the plate. There is a good reason James Beard Award–winning writer and video producer Michael Gebert named his Midwest-focused video series "Sky Full of Bacon"—there is a lot of pig here. Gebert's videos explore far more than pigs, but the importance of pigs in the region makes them a dominant presence on farms and in kitchens, and therefore a regular focus of his reportage. He reveals the trends and cultural influences that have made Chicago, as the opening credits state, "a city with a sky full of bacon," but also tracks trends and foods to surrounding states. From his in-depth portrait of Hispanics making *carnitas* to the importance of barbecue among African Americans to the rising interest in heritage breeds, his videos make one thing clear: pigs are more than just food—they're culture.

Previous chapters have mentioned groups that settled the Midwest, but it is worth noting that the deeply entrenched pork traditions brought from Europe, Asia, and Latin America often remained best known within the cultural milieus that cherished them. However, as more and more diners have become interested in exploring a wider range of cuisines, it was inevitable that they would be increasingly eating pork. Appreciation of pork in the Midwest stretches back a long way, but we continue to discover new ways to enjoy it.

Celebratory Pig

State fairs appeared in the mid-1800s, enthusiastically encouraged by Abraham Lincoln. While fairs celebrated more than pigs, pigs were certainly there in substantial numbers. People were developing new breeds to meet the needs and demands of the day, and improvements in all aspects of farming were definitely what Lincoln hoped for. As Lincoln stated during a speech given at the Wisconsin State Fair in Milwaukee on September 30, 1859,

> Agricultural fairs are becoming an institution of the country; they are useful in more ways than one; they bring us together, and thereby make us better

> acquainted, and better friends than we otherwise would be. . . . But the chief use of Agricultural Fairs is to aid in improving the great calling of *Agriculture*, in all its departments. . . . And not only to bring together, and to impart all which has been accidentally discovered and invented upon ordinary motive; but by exciting emulation, for premiums, and for the pride and honor of success—of triumph, in some sort—to stimulate that discovery and invention into extraordinary activity.[1]

In other words, right from the outset, the state fair was friendly, informative, and designed to celebrate, reward, and encourage excellence.

Today, state fairs still bring people together for many of the same reasons, though improved communication technology makes it less important as a means of keeping up with the latest developments. However, the celebratory aspect and the rewards for excellence are still there. Eager youngsters guide the handsome porkers they've raised around the show ring, hoping for a prize. Breeders bring in their stud boars, looking not only for blue ribbons but also new customers. While all things agricultural are on display, livestock exhibits take up a substantial part of any fairgrounds. When pigs are important in a state, they will be important at the fair.

Not too surprisingly, pork is also available for consumption. This will almost always include corn dogs, perhaps the most universal of American fair foods on sticks. Hot dogs, chocolate-covered bacon, and barbecue appear at multiple fairs. In Wisconsin, fabulous bratwurst is abundant. Italian and Polish sausages are on hand at Illinois's fair. Minnesota has deep-fried baby back ribs, as well as (of course) SPAM burgers. Nebraska has double-bacon corn dogs, bacon-wrapped meatballs (called "moinks"), fried SPAM on a stick, and plenty of smoked pork. Traditional at the Indiana State Fair is Sati-Babi, marinated pork on skewers, first created three decades ago by a Filipino entrepreneur in Terre Haute.

Iowa, being pork central for the nation, does an especially good job of making sure pork is available to eat. The Iowa Pork Producers Association (IPPA) has a permanent building on the fairgrounds, where a tent-covered dining area led to its being dubbed the Iowa Pork Tent. IPPA member Ron Birkenholz relates,

> At the Pork Tent, we sell Iowa Chops, pork burgers, pulled-pork, and pork loin sandwiches. In addition to the tent, we have two trailers selling Pork Chops on a Stick. The IPPA introduced the Pork Chop on a Stick at the state fair in 1999, registering the trademark, as we wanted to keep it Iowan. Though foods on sticks are popular at fairs, this isn't really on a stick. It's a 9 to 11 oz. frenched pork rib chop. When you french a chop, you keep the bone attached

> to the meat but clean off the bone, so it can act as a handle. That way, people can eat the pork chop without utensils. The Pork Chop on a Stick, which is marinated and baked, regularly gets named "best fair food," and it's our biggest selling item. We average more than 60,000 sold each year over the course of the 11-day event. This year [2015], we had sold our entire supply of 65,400 chops a few hours before the fair ended. Fortunately, even though the Pork Tent doesn't have Chops on a Stick, they still had food, so everyone could eat right up to the end of the fair.

In the late 1800s, people in the Midwest decided that something permanent and more focused than a fair was needed to celebrate the pig. In 1887, Sioux City, Iowa, created the world's first Corn Palace, announcing that they were honoring corn because it had made Iowa the dominant force in the region's pork market.[2] Even today, in the Mitchell Corn Palace, in South Dakota, there is a display that reminds us that pigs were once "corn on the hoof" and relates that a bushel of corn, when consumed by a pig, is converted into roughly eight pounds of pork.

In areas where raising pigs is important to the local economy, people began to plan local celebrations. In the mid-1900s, pork festivals started appearing across the Midwest. Often, these are held at the county or town level, with multiple events in each state. All these events offer food in abundance. Some have sash-wearing pork "royalty." Some focus on the economics of pork or best practices in pig raising. Others are just an opportunity to eat pork.

The Preble County Pork Festival in Eaton, Ohio, is one of several pork-related festivals in Ohio. Launched in 1971, it has grown from a dream to an event that draws more than one hundred thousand people a day. There are demonstrations of pork cooking, butchering, and sausage making. Fairgoers can view pig races and a sow with her litter. Children can display their pig art and attend a Just for Kids Farm Safety class. The queen and king are selected from among high school students who have livestock projects entered in the fair. So, all ages are involved. Then, of course, there's the food, starting with all-you-can-eat pancakes and pork sausage for breakfast, and sustaining people throughout the day with pulled pork (sandwiches or nachos), ham, smoked sausage, and a barbecue pork chop smorgasbord.[3]

The Tipton County Pork Festival, in Tipton, Indiana, got started in 1969. Visitors can "pig out" on pork chops, pork burgers, breaded pork tenderloin, and more. There is a Miss Tipton County Pork Festival Queen Pageant, but also a Princess competition for girls aged six to nine years old.[4]

The C.H. Moore Homestead DeWitt County Museum in Clinton, Illinois, focuses more on history at its Apple 'n' Pork Festival, with demonstrations of

Photo 11.1. People line up to buy Pork Chops on a Stick at the Iowa State Fair. Photo courtesy of the Iowa Pork Producers Association.

crafts from the 1800s and a Civil War encampment. Here, the menu includes ham and bean soup, smoked ham sandwiches, and grilled bratwurst.[5]

These are just a few of the pork festivals in the Midwest. There are many more. In addition, there are "Porktoberfest" events across the country, including one in Minneapolis, where beer is (not too surprisingly) part of the focus, along with the requisite pork-based dishes (brats, barbecue, and Korean pork belly sliders).

The more urban a festival location is, the more likely it will focus on eating, rather than on livestock. It will also be more likely to focus on a single pork product or application, rather than on pork in general. There are festivals all across the region for sausages, side pork, bratwurst, ribs, and grilled, smoked, or barbecued pork. There's even a Pork Rind Heritage Festival in Harrod, Ohio. If something is appreciated and enjoyed, there is probably a festival.

Among those pork applications that have become the focus of obsession, few have generated more enthusiastic devotion than bacon. Traditional dishes, such as bacon-wrapped dates, have staged a comeback. People have created dishes with names like "bacon bomb" and "bacon explosion." Different styles of bacon have been developed. Bacon is almost literally on every lip.

The most intensely bacony of all celebrations is also the newest: bacon festivals. Across the Midwest, and even across the country, large-scale bacon-focused events have increased in number in recent years. Seth Zurer, co-founder of Baconfest Chicago, has thought a lot about why bacon is so popular, as well as about how long this obsession is likely to last. "Bacon has an archetypal flavor. It evokes memories of a less complicated past, when we ate bacon with our families at breakfast," Zurer observes. Plus, it tastes incredible. That evocative flavor helps draw people in, and it was the taste more than anything else that led us to start Baconfest." Zurer and his two partners actually have theater backgrounds.

> We saw a rock and roll puppet musical about beer, and we thought to ourselves, "what could we write about that we love as much as beer?" and the immediate answer was "bacon." However, we thought it didn't lend itself to theater as much as to a party. So, in 2009, we spent six months raising support for Chicago's first Baconfest. I'd done some writing for food magazines, so I had some contacts. We never envisioned it as being solely a party—we wanted to use it to raise money for the Food Bank of Chicago. Part of the proceeds are donated, plus we ask people who attend to bring nonperishable foods to the event.

Zurer also notes that the economy has had something to do with their success. "In 2009, when we started the event, there was a financial crisis. Bacon was an affordable luxury. That first year, we had 75 guests in attendance, but by the next year, we were up to 600 guests. In 2015, we had 4,500 tickets, and we sold out."

While the flavor of bacon and the chance to sample original dishes created by top chefs using bacon draws consumers to Baconfest, it's the charitable aspect that brings in the big talent. "It is hugely important to area chefs that we end hunger," Zurer emphasizes, "and that's the main thing that draws them to the event. This year, we had chefs from 170 restaurants donate their time to this project, and Nueske's Applewood Smoked Meats donated 8,000 pounds of bacon."

At the heart of Baconfest is a competition. Each chef is given fifty pounds of bacon to create a dish, and the results are judged by food writers, restaurant critics, and other food professionals—and sampled by the thousands attending the event. "These are top chefs," Zurer notes, "James Beard Award winners and chefs with Michelin stars, as well as hot up-and-comers. The competition generates an amazing array of wildly different creations. These chefs understand the deep emotional connections most of us have with bacon, and they know how to tap into that."

There are two categories in the pursuit of prizes: Most Creative Use of Bacon and Best Front of House Presentation. The prize is the Golden Rasher Award, a small golden pig statue atop a wooden stand. There are also prizes for the "bacon public," including a competition for best bacon poem.

Though the goal is to end hunger, there is also a great deal of fun involved. "We're experiencing kind of a 'Golden Age of Smoked Meats' in Chicago now," Zurer observes.

> As has been traditional with so much of pork cookery over the centuries, this event elevates an essentially humble ingredient and makes it fabulous gourmet fare. We found that the chefs have a deep passion for it and enjoy using it as an ingredient. Plus, all the people who attend the event get to taste this amazing food. So, everyone is having fun.

Given that the tickets, which now range from one hundred to two hundred dollars, sell out within hours of going on sale, it seems clear that Zurer and company have definitely tapped into what people want. "People have three hours to sample as many of the remarkable bacon creations as they can. It's definitely an indulgence. But in 2015, we donated $75,000 and 1,768 pounds of food to feed the hungry. I feel that that is a karmic counterweight."

Participatory Pig

Today, an increasing number of folks want to go beyond simply indulging in food. They want to be involved in the process. This can take the form of anything from learning how to cure meat to "adopting" a pig, watching it grow, and having it end up in one's freezer to participating in the increasingly popular barbecue competitions. Whatever outlet is found, people seem to want to either know where their food comes from or take part in producing it—and sometimes both.

Matthew Zatkoff slices and carefully arranges a variety of sausages, including a particularly nice *finocchiona* salami, pancetta, and impressively dark country ham on a heavy wooden cutting board. Nearby, bacon sizzles in a small frying pan. While he works, Zatkoff chats about filming and editing videos for clients—because creating splendid charcuterie is not his day job. "Once I get interested in something, I want to know how it's done," Zatkoff explains. How he got interested was when a friend from Virginia told him about country ham, as well as about legendary country ham- and bacon-maker Allan Benton.

Zatkoff was intrigued. He ordered one of Benton's country hams, and he was hooked. That's when he started to do research.

> I went online and started finding out how meat had been cured through history. People have been doing this for thousands of years. I looked through archives for old techniques, recipes, advice. Even old newspapers often had recipes for curing meat. I wanted to see how folks used to do it. That's how I cook anything: research the history first, comparing old techniques with new. A ham was the first thing I cured. Then I did another ham, and then bacon, then a few styles of sausage. Fortunately, once meat is cured, it lasts just about forever, so you don't have to worry about making too much. After the meat reaches the desired age, I sometimes vacuum pack it, but that's just to preserve the flavor and texture. It's the salt, smoke, and years of aging that preserve the meat.

Zatkoff also ordered more country hams. He has now tried at least half a dozen different ones, comparing flavor profiles and consistency. "One ham wasn't very interesting, and I wondered if it would benefit from more aging. So, I hung it up and kept it for another two years, and that really improved it."

Knowing others who also love food helps, as Zatkoff and friends have on a number of occasions bought whole hogs and divided them up. "We've done both heritage breeds and regular pigs. I wanted to know how different pigs reacted to curing and smoking. The most interesting breed I've used so far was probably the Iowa Swabian Hall. The fat content was remarkable."

Photo 11.2. Homemade charcuterie can be impressively professional looking, as this presentation of Matthew Zatkoff's handiwork illustrates. Photo by Cynthia Clampitt.

Zatkoff relates that this is definitely a hobby that requires hope and patience.

> You're taking a risk when you hang something up and have to wait three years to discover if it turned out. I actually think that's kind of exciting. Fortunately, I haven't had a bad batch yet. Of course, the research helps here. If you've done everything right, it's a calculated risk. But it's not just making it that's

fun. I love that it gives me something to share with friends. However, I always try it first. If I don't get sick, then I'll share it.

Asked what advice he'd give to people who want to try this themselves, Zatkoff of course recommends research, not just on recipes but on equipment. "There is a lot of equipment out there that would help people who want to limit the risks, but the price might scare people away from getting interested in the hobby." Zatkoff also wants people to know they don't have to make a huge investment to try this.

> Some people think you can't cure meat without temperature- and humidity-controlled facilities. You might need them if you're going to have a commercial operation, but not if you're doing it for fun. Throughout history, people have used spare rooms, barns, and tool sheds, and they've cured meat in a wide range of climates, all over the world. You don't need to spend thousands of dollars. I use a spare room for hanging the hams. If it gets dry, I turn on the humidifier. If it gets hot, I'll crack open a window. Paying attention is more important than fancy equipment.

Zatkoff relates that food has always been an interest—and he has always liked experimenting.

> I grew up making sausage with my family at the holidays. Also, when I was a kid, my brother and sister and I would play restaurant. We'd go into the kitchen and pick random ingredients, then challenge each other to create something with them. Back in Indiana, I worked for a while in a sushi restaurant, but it was the staff meals prepared by the Korean owners that interested me even more, as that was less familiar. Then, when I moved to Chicago, one thing I'd do for fun was hop on the subway and get off at random stops and just look to see if they had any interesting restaurants. I don't know how safe that was, but I really discovered a lot.

That sense of exploration clearly serves Zatkoff well in the culinary world. "I love it all—all the cured, smoked meats—but I'd like to get more into different types of sausages. There are so many varieties. Just about every culture in the world has sausage. There's a lot out there still to explore."

Oak Park–based food writer David Hammond thought that the rare and unusual American pig breed known as the Mulefoot would be an interesting gift for his wife for their wedding anniversary. A live one. Fortunately, his wife is a good sport, plus she enjoys food as much as he does. The Mulefoot is a heritage breed that is unusual in not having a cloven hoof—its foot looks more like that of a mule. David explains, "A number of food writers in the area were studying these pigs, and I became intrigued. I wondered if breed really made that much of a difference. We have a freezer in the garage, and we'd been playing with the idea of getting a quantity of meat to fill it, and I figured this was a good way to accomplish that."

Hammond relates that the farmer from whom he bought the pig goes through the alphabet a year at a time, with all names in any year starting with that year's letter. When Hammond made his purchase, the letter was E. The first name that came to mind was Ermine. "It was my grandmother's name. It may seem odd, but I really don't think she'd mind."

Ermine's home in Argyle, Wisconsin, is less than three hours from Hammond's home.

> We'd go up every six weeks or so, taking food from the garden as a gift for Ermine. Ermine and the other pigs seemed to be genuinely happy. They lived in wonderful conditions, or so it seemed to us. There was lots of mud for them to wallow in, and the scenery was lovely. The emotional connection was interesting, too. Ermine enjoyed the food we brought, but she clearly didn't like me.

Hammond relates that the price was fairly reasonable. One pays for the pig and, assuming one doesn't want to kill and butcher one's own pig, for the processing. "The pork we got from Ermine was outstanding. The fat was fluffy, and the pork chops were outrageously fatty. They were just fabulous. This was our way of putting into practice the theory of paying more and eating less but better quality."

While many folks simply attempt barbecue at home, happily enjoying the results with friends and family, an increasing number of people are getting involved in barbecue competitions. And in the world of competitive barbecue, few people are more involved than Carolyn Wells. Wells is co-founder and executive director of the Kansas City Barbecue Society (KCBS), the world's largest organization of barbecue enthusiasts. KCBS has grown from thirty people in 1985 to more than twenty thousand members today.[6]

Wells has a great smile, a friendly, easy-going manner, and a wonderfully smoky voice that suits her calling. She relates that, being a daughter of the South (she grew up in Nashville, Tennessee), barbecue was always part of the culture. Her grandfather was a country doctor who traveled on horseback and accepted food as payment, and it was he who introduced Wells to barbecue, usually at church socials hosted by him and Wells's grandmother. "To me, barbecue has always meant food, family, fun, and friends," Wells relates. "Barbecue crosses all economic lines. If you love people, you feed them, and there is no better way to feed them than with barbecue."

During summer visits to her grandparents' farm, Wells learned to appreciate the value of pigs, as well as of barbecue.

> A pig is useful because you can use everything. When I was a child, Mom and Dad would pick a cold day in February, and we'd all gather and slaughter a pig. Everyone worked together. The pig would be butchered, salted, smoked, we'd stuff sausages—everything. Meat was available in stores, but buying from a store was an exception. All of us on farms processed our own pigs. We even got out of school for this event.

Speaking of her start in the world of competitive barbecue, Wells relates that she once worked for a group of lawyers who owned a barbecue sauce company. The company's major market was Memphis. So perhaps it was inevitable that she'd be drawn to Memphis in May, a month-long festival in Tennessee that is home to, among other things, the World Championship Barbecue Cooking Contest, one of the top contests in the country. She and husband Gary moved to Kansas City in 1976 but continued to pursue barbecue at top events. Then, in 1982, they entered the American Royal Barbecue competition, located in their hometown of Kansas City. "All we cooked was ribs, but we won—and I was hooked."

At that time, competitive barbecue was all done at the local level, with different rules for every competition. Wells saw a need, and she, her husband, and friend Rick Welch decided to meet that need, forming a society that would help level the playing field for competitors. "We needed guidelines, needed to codify expectations, make it fair. We were approached almost immediately by someone who wanted to have a competition and wanted it to be sanctioned. That's when we began to formalize the rules and standards. Being a new sport, we all grew up together." Wells, now a widow, notes that since they started thirty years ago, the number of barbecue contests has grown dramatically. "Today, KCBS sanctions more than 500 competitions annually."

Wells emphasizes the emotional side of the competitions, not just the great food. "Barbecue, even when people are competing, is all about community. Barbecue people are belongers and givers. But they're also strong characters. Definitely individuals, but very community oriented. They think it's their duty to give back. They also tend to like tradition." Wells mentions Operation Barbecue Relief as an example of the kind of people attracted to barbecue. Operation Barbecue Relief is an organization of volunteers who move their big smoking rigs to the sites of disasters, to serve good barbecue to both emergency crews and the people devastated by tornadoes, floods, and other catastrophes. "People who are involved in Operation Barbecue Relief will tell you that they get so much more than they give."

Wells continues,

> I think this sense of community is one of the things that has made barbecue increasingly popular in recent years. On 9/11, I was serving on a federal grand jury. After the attacks, the jury was cancelled and we were sent home. There was a contest scheduled that weekend, and we thought it might be cancelled too, but we wanted to go. We drove to Decatur, Alabama, to Riverfest. Such a beautiful place. And what we found was that everyone felt broken and they wanted to be with other people. They wanted a sense of normalcy. So, the turnout was bigger than ever. We saw a fifteen percent annual growth after 9/11. People wanted comfort, and barbecue is the ultimate comfort food. Plus, barbecue is such a strong part of Americana. People wanted to feel American then, too.

Wells also notes that barbecue is a great antidote to the rushing/bustling lifestyles that most people lead these days.

> Low and slow is how we cook, but it's also the attitude we bring to barbecue. It's not a solitary pursuit. Even in competition, it's a team effort. Small- and medium-town America decided barbecue was the place to get everyone together. The media has helped in recent years, especially with shows like *Barbecue Pitmasters*. That drove a huge increase in interest. Barbecue has become mainstream enough to be valued. It used to be kind of looked down on, but now it's the hot new old food.

There are now several sanctioning groups, but barbecue has remained friendly. "We feel we can cook under anyone's rules and still have it be fair." Wells notes that competitive barbecue is still a hobby for most, but now that there is prize money, a handful of people do manage to make a living at it.

> For some people, a win at an individual competition is all they want. However, some barbecue teams hope to win Team of the Year. For those pursuing

Photo 11.3. Smokers stand ready at the KCBS-sanctioned Red, White, & Bar-B-Q competition in Westmont, Illinois. Photo by Cynthia Clampitt.

> that honor, competing in 35 to 40 contests is common, and some competing teams go out 42 weekends of the year. Points are figured based on your ten best contests that year, and the points chase can become pretty heated by the end of the season.
>
> There are five elements in barbecue: the smoker, the fuel (charcoal or wood are all we allow), the protein, the seasoning, and the expertise of the chef. Everyone tinkers all the time with the seasoning and the techniques, trying to refine them. Flavor profiles have become so much more sophisticated than they were when we started. Go back East, and the barbecue focuses on pigs. They don't really think it's barbecue if it isn't a pig. In Memphis, the categories are whole hog, shoulder, and ribs. In Kansas City, while two of our required categories are pork, we also recognize brisket and chicken.

Barbecue keeps Wells busy. She is the author of *Barbecue Greats: Memphis Style* and *Memphis Barbecue, Barbeque, Bar-B-Que, B-B-Q*, and co-author, along with Paul Kirk and Ardie Davis, of the *Kansas City Barbeque Society Cookbook, 25th Anniversary Edition*. And competitions can take her far afield. "There are 60 contests in a 100-mile radius of Kansas City, and even the World Pork Championship in Washington, D.C. is KCBS sanctioned. There are now contests from coast to coast, and even interest overseas. But I still go back every year to where I started, to Memphis in May, but now I go as a judge."

Barbecue is not the only pig-focused competition, though it is the only one that is almost entirely amateurs. For professionals, there are, among other opportunities, the Taste of Elegance, state and national competitions sponsored by the Pork Board, and Cochon 555, regional and national competitions that spotlight heritage pig breeds and promote the family farming movement.[7]

Getting a backyard smoker, buying a pig to fill the freezer, or forming a team to compete for barbecue glory are just a few of the ways one can participate in making the most of pork. Of course, for the majority of folks, the primary way they participate is enjoying home cooking—or dining out.

Exceptional Pig

In recent decades, upscale restaurants have been not only exploring new options with pork, they have also been focusing on making everything about the dining experience exceptional. In addition, specialty retailers have been seeking out producers who are raising meat sustainably or organically. Prices for these are higher, but taste is the driving force, not economy. Among the hot trends today are nose-to-tail dining (everything old is new again) and heritage breeds (another back-to-the-future item), especially when prepared by remarkable, even visionary, chefs. Chefs like Paul Kahan.

Paul Kahan, the James Beard Award–winning chef who, with partner Donnie Madia, rules a veritable empire of critically acclaimed dining establishments, wears a perpetual look of barely suppressed laughter. One easily imagines that he enjoys what he is doing. He greets staff members, checks in with his assistant, and pops into Madia's office as he walks through One Off Hospitality, the management group that coordinates the eight (at present) venues for which Kahan is executive chef. The offices, where the air is perfumed by the on-site bakery, are within a block of two of Kahan's restaurants: The Publican and, across the street, Publican Quality Meats, a café with an attached butcher shop.

Kahan didn't start his career in the kitchen. In college, he studied math and computer science. But his computer job wasn't fulfilling, so about thirty years ago, he quit to pursue becoming a chef. "I've been in the kitchen ever since," he states. "When I started this new career, I made a decision to have a high level of ethics and to take care of farmers and employees. That's paramount: take care of and respect everyone." That may be the most important goal, but next in line was making really amazing food. Kahan notes,

> Pig is a big part of just about every culture in the world, and we're tapping into that. Growing interest in pigs has paralleled growth in American cuisine. Knowledge of how food is produced is spreading. What people want is changing. When I worked with Rick Bayless at Topolobampo, a whole roasted pork loin was a big deal. Now people are more interested in other parts of the pig. At avec, we have a wonderful braised pork shoulder. However, I don't see what is happening as something new, but rather as a resurgence of ideas that have been neglected.

Today's diners are not just looking for different parts of the pig; they're looking for different pigs. "I want to cook one of each type of pig," Kahan relates, "to find out which one is best for each application. For example, for what we do, Dutch pork is the best choice for pork belly dishes."

Where the pigs come from matters a lot to Kahan. He knows the people who raise the animals served in his restaurants, he knows what each farmer has to offer, and he knows their level of devotion to animal welfare. He rarely mentions a type of pig without identifying who raised it. "At Blackbird, we bought suckling pigs, Mulefoots, from George Rasmussen at Swan Creek Farm in Michigan. We would *confit* them in duck and pork fat." At The Publican, the menu identifies the farms where each item of food was purchased, but even in conversation, Kahan eagerly shares the names of his sources. He likes the organic Berkshire/Chester White cross from Jude Becker in Iowa, and praises the Berkshire/Duroc crosses from Louis Slagel's Illinois farm. Kahan relates,

> The Publican brand has really grown. The whole concept was based around my twin loves of pork and beer. The restaurant did well, and now we have Publican Quality Meats. In the butcher shop, we have pork from five different farmers. It's all beautiful pork from heritage breeds. We began curing our own meat when we opened avec, and we still do that here. At PQM [Publican Quality Meats] we make 12 or 14 kinds of sausage. We also have someone in Wisconsin producing sausage to our specifications. But all the meat at PQM is amazing. I think our porchetta is the best out there. We roast a whole porchetta every Saturday morning, and when we open at noon, people line up to get some while it's still hot. It's not as good the next day.

Kahan doesn't just love buying and selling pork. "When I go ice fishing up in Wisconsin, there is a place my friends and I stop and buy three or four pounds of bacon. The whole time we're fishing, there's a skillet nearby with bacon sizzling." Plus, Kahan hosted the first Baconfest at The Publican in 2009. "Bacon is big because it's delicious. It hits all the high points: salt, fat, smoke. It's super craveable. My favorite bacon is *guanciale*, which is made from the pig's jowls. It's so complex in flavor. That said, all bacon is good."

Interestingly, though Kahan didn't start his career in the kitchen, he did start his life involved with food. He was born in Chicago's South Shore area, and his family was in the food business. "Dad had a delicatessen, catering to the Jewish market. I worked there when I was young. Later, dad had a smoked fish business in Chicago, King Salmon, also for the Jewish market. I drove the delivery cart. The way I related to my dad was going out to eat with him. As a result, I've always been intrigued by food and how it connects people."

Kahan does not shy away from talking about being a nice Jewish boy who grew up to be closely associated with pork. However, he also acknowledges that he is far from being alone. There are enough others, in fact, that he jokes about starting a club.

Kahan clearly appreciates the culinary range of pigs, noting that his favorite cut is the gnarly, fatty pork chops near the neck, though his favorite dish on The Publican's menu is the country rib preparation. "The ribs come from the Slagel Family Farm, and they're wonderfully flavorful." Clearly, what Kahan likes is what diners like, as the success of his restaurants keeps growing. "Next year," Kahan says happily, "we'll crack the 1,000-employee mark."

A block away, I step into Publican Quality Meats, where employees happily shared their thoughts about PQM, pork, and Paul Kahan. Apparently, he's every bit as good a boss as the press suggests. "I work here because I get to work with Paul," enthuses one employee. When asked for sausage recommendations, one young man expresses doubt that any one sausage is better than the others, but then recommends the *boudin blanc*. In the refrigerator case, a package of scrapple also beckons, as do the fatty country ribs in marinade that Kahan says he favors. The prices are high, but that is the case with all truly remarkable food. This is not a place to shop if one is pinching pennies. It is a place to shop if one wants to experience exceptional pork prepared exceptionally well.

Is it worth it? That will depend on the consumer. There is a notable difference between heritage pork and ordinary pork. The flavor of heritage pork is bigger. The fat is softer and far more abundant, though in heritage pigs, having lots of fat is part of the attraction. But, depending on what one is buying and where one is buying it, prices can run anywhere from two to ten times higher than regular grocery store pork. It is wonderful that pork this fine is available. In fact, it's important, as having many breeds of pigs available helps protect genetic diversity. And having choices is good. It is also wonderful that pork is available that is affordable, because most consumers cannot afford the high-end meat, at least not regularly. (I certainly can't.) Fortunately, even quite ordinary pork is delicious when prepared well.

CHAPTER TWELVE

Farm to Table Pig

People Who Raise Pigs and Create Pork

Pigs and people have been living together for roughly twelve thousand years. A lot of things have changed in that time, including where pigs live. Today, in the Midwest, raising pigs and getting them to market and then the dinner table are different from even a hundred years ago. However, the need to have people available to make these things happen has not changed. In fact, because of the extra layers of work, including animal welfare considerations and food safety inspections, there are more levels than ever at which people need to be involved. No more letting pigs forage in forests or alleys and getting rounded up by a swineherd as winter approaches. At least not in the United States.

Of those many layers of people, I don't think it is possible to find anyone who doesn't take his or her job seriously, and yet they all seem to find joy in doing it. Meeting these people was delightful and encouraging.

Bringing up Piggy

The size of a farm has nothing to do with the level of commitment for the farmer. Small farms can offer the option of having a greater variety of animals or needing less start-up capital. Large farms are needed in order to feed growing populations. Having multiple approaches to raising the animals means both farmers and consumers have choices. We'll look at both, but we'll start small.

Judi Weiss, a farm girl from North Dakota, moved to Minnesota to join her then fiancé, Josh, at Back Home Farms. They knew they would want to diversify eventually, but the couple decided that raising pigs would be the best place to start. "We did some research on heritage hogs," Weiss relates, "as we didn't want to raise the same old pink pigs everyone else was raising." The Weisses decided to focus on Berkshire crosses. "We now have two boars that are purebred Berkshire and one that is half Berkshire. Those three, along with our fourteen breeding sows, are about all we can handle, since there are only two of us running the place—plus we've added chickens, turkeys, and sheep since we started."

Weiss and her husband fell in love with their pigs. "All our sows and boars have names, and we greet them by name every time we see them. It's funny how they learn their names so fast. They each have a personality and will actually interact with you. We even have a sow that will fetch a stick."

The farm's sows farrow twice a year. "We understand why some people use crates for farrowing, but we didn't want to go that route. We farrow in open pens, with smaller pens attached for the babies to get under a heat lamp. Of the pigs that are born here, we sell many as feeder pigs—that is, weaned youngsters that someone else will feed up to full weight. We also have a nice list of butcher customers, where we raise the pigs to full market weight before selling them."

Weiss relates that the farm is definitely a joint venture, but she feels that working together on the farm strengthens her relationship with her husband. "All decisions are made together, whether it's switching pens around, when to breed, when to wean, when to cull, whatever. We are both involved in every aspect, from grinding the feed for the pigs to castrating to caring for sick pigs."

That said, Weiss acknowledges that she does more of the marketing, both of breeding stock and the butcher pigs. She enjoys this, as it gives her a chance to talk about the pigs. "I love telling people how they act, what garden produce they love best, how we take care of them, what their habits are.

"We have grown a nice clientele for both our feeder pigs and butcher hogs. We really focus on the 'Berkshire difference,' and our customers are hooked once they taste that difference. Our motto is 'Know your Farmer, Know your Food.' We're not organic, but we don't use antibiotics, and we grind our own feed. Our pigs get the best care we can provide."

Weiss acknowledges that one has to be careful with pigs. "As you know, pigs can be vicious, and they are omnivores. We don't feed our pigs meat or meat byproducts, and we hope this makes them less of a threat. We also interact with our pigs as much as possible. We can go into the pens with our boars and even our sows after they have babies. We know to never trust them a hundred percent, but by raising most of our pigs from babies, we gain their trust and respect."

One of their biggest concerns is the weather. "In Minnesota, winter is definitely the hardest time to have pigs. The heritage pigs are pretty tough, but keeping their water troughs clear of ice and making sure there is enough space in the shelters, so they can get out of the cold, can be a challenge. In the cold, there is also the danger of huddling hogs crushing the little ones. So, we have to watch the pigs closely during the winter."

When talking about the pigs they butcher for their own use, Weiss employs the same "use it all" tradition that has been a thread through so much of history. "The head usually goes to my husband's grandmother, for making head cheese. Josh and I love the pig heart. Our chickens love to peck the skin. I render quite a bit of lard to use in cooking and in baking—and my buttermilk drop donuts fried in pig lard are very popular. And, of course, we have a lot of great pork to eat!

"I work from a home office, and my window overlooks a couple of our pig runs. I feel so lucky that I get to watch our pigs every day."

At the opposite end of the size spectrum, but feeling equally fortunate, is Malcolm DeKryger, who sees his work as a calling and an opportunity to practice good stewardship—a feeling he says is shared by most of the company's employees.

DeKryger is president of Belstra Milling, a family- and employee-owned business in northwest Indiana. In addition to producing feed for a variety of animals, Belstra also operates six pig farms that house 14,200 sows and produce nearly four hundred thousand pigs a year.

"It takes enormous passion and dedication to work with pigs," DeKryger notes. "It's satisfying work, but not easy work. You're moving all day, and you have to be constantly thinking about feed, health, facilities, and all the other aspects of pig raising."

Clearly, DeKryger has that passion. The books, diplomas, and certificates that fill his office testify to that, as does the bronze pig statue that bears the famous quote from Winston Churchill: "I like pigs. Dogs look up to us. Cats look down on us. Pigs treat us as equals."

In addition to passion, it also takes a remarkable support network to handle the job well. DeKryger explains that there are five general categories in pig production. All are vital, and all are constantly changing. The five are feed/nutrition, genetics, health, facilities, and management.

"In the category of food, we have great technologies," DeKryger relates, "but there are variables we can't control, such as the weather. For example, this year we had too much rain, and three-quarters of the corn crop was lost on some farms. You have to be able to find alternative sources in those cases." Belstra needs three million bushels of corn per year for their pigs. Of course, corn is not the only thing pigs eat, and nutritionists make certain the blend of grains and nutrients will keep the pigs healthy. But corn is a key ingredient in that feed, so obtaining enough is vital.

Genetics is a constantly evolving item in the farmer's arsenal of tools for raising pigs. "This field is still being fine-tuned, but it's proving helpful," DeKryger acknowledges. "The genes on DNA strands are being tracked. We can identify the traits we want, check the DNA, and pick pigs that have the genes we want. It really makes breeding more reliable."

Now that "fat fear" is diminishing, the genetics company with whom Belstra works is breeding for more interstitial fat, also known as marbling, those slender veins of fat and flavor that run through meat. Pigs can also be bred for more Omega 3 fatty acids, the kind one expects from fish. DeKryger emphasizes, "This is not genetic modification; some pig breeds simply have more Omega 3s. Use those pigs when breeding, and you end up with more Omega 3s in all your meat."

While all the categories are important, health is probably the one that causes the most concern. Of course, there are the day-to-day tasks handled by local veterinarians. However, having large numbers of pigs in close proximity to one another means disease can spread quickly, so bio-security is continually reviewed and improved. "Last year, there was a devastating epidemic of PED [porcine epidemic diarrhea]. The U.S. industry lost between eight and ten percent of all baby pigs. We have to be careful to not panic and start pointing fingers, and it's not entirely certain how it entered the U.S., but the diagnostic center at Kansas State used the phrase 'Chinese footprint.' That doesn't mean it was intentional, but it's a likely source of the virus. So, we've tightened up bio-security measures.[1] Fortunately, we have better tools all the time. Diagnosis and treatment are constantly improving. We can find solutions faster than ever before."

Research is not limited to health. Facilities are constantly being examined, modified, and upgraded. Animal welfare experts across the country and around the world are always looking for new ways to reduce stress among animals and improve their general well-being, and a lot of that has to do with how and where they live. DeKryger explains, "We are always looking for the 'next best thing,' the newest insights, best designs, most efficient structures, whatever is needed to keep the pigs safe, healthy, and happy. That said, there are always tradeoffs, and no situation is completely free of problems. Sows now have more room to move around and interact, which they like, but it also means more fighting and injuries. There are fewer injuries with stalls, but the sows can't move around. It's not always an easy decision. The meat is the same in both cases. But on the whole, the pigs like interacting, so we give them space."

Management gets back to that "enormous passion." Everything going on in the other categories has to be orchestrated. Technology offers assistance with this. "Our IT folks give us more tools to help us manage the operation. Computers help us improve consistency. They maximize what can be done by tracking everything that happens. They make it possible to treat each pig in more specific ways." DeKryger's son Nick, who has a master's degree in agricultural economics, is also part of the management team.

"We can do so much here in the Midwest, because we have the space. We have the crops that feed the pigs and the pigs produce the fertilizer for the next crop of food. It's a truly sustainable recycling system. We are continually improving this process, and we've been able to dramatically reduce our carbon footprint," DeKryger concludes.

As much as he values the pigs, the thing that really keeps DeKryger going is the fact that he can feed half a million people three quarter-pound servings of protein per week, every week, year-round, from just one of the farms—and do it with a carbon footprint that is vastly lower than if people had their own pigs. Feeding people. Protecting the environment. Good stewardship, indeed.

As noted, it takes more than just farmers to keep the world of modern farming moving forward. University extension programs send out representatives to provide the latest research to farmers. Scientists study health issues. Animal behavior specialists fine-tune enclosures or transportation, to reduce stress in animals. Co-ops and family-owned businesses, such as Sioux Nation Ag Center or Pipestone Systems, provide everything from veterinary services to nutritional specialists to genetic research to marketing the grown animals.

Tim Heiberger, who is a field marketer for Sioux Nation Ag, knows not only the names of all the farmers in the area he serves, but also what kinds of barns they have, what they raise, and how they're doing it. Driving down the long, straight, dirt roads of an iconically level section of South Dakota, he points out the key features of the handsome farms we pass, relating what each device or refinement accomplishes, from making animals more comfortable to reducing the risk of disease spreading. "I'm impressed with how efficient and clean the process has become," he states emphatically. The wean-to-finish barns are a relatively new invention, but they are being adopted rapidly. Heiberger feels certain that, if people understood how much protein the world is going to need in the future, they'd show greater interest in efforts to improve protein production. Typical of the friendly generosity of the Midwest, Heiberger not only talks about Sioux Nation Ag Center, but acknowledges that competitors do a great job, as well. But then everyone out here, in rural America, knows that they're all in this together.

I've always loved visiting farms, but increased bio-security in recent years makes such visits increasingly difficult, at least for the big confinement operations. Even people who work on these farms generally have to shower and change their clothes and shoes before coming into contact with the pigs, to avoid accidentally bringing in outside contaminants. However, for those who want to see what a modern pig farm is like, there is The Pig Adventure at Fair Oaks Farms in Fair Oaks, Indiana. The families that own Fair Oaks Farms are working to reconnect consumers with the sources of their food, while also demonstrating best practices in animal welfare, environmental concern, and sustainability. This newest addition to that project gives visitors the chance to witness a large-scale operation without any risk to the pigs. The pigs remain safe for the same reason a patient remains safe in an operating theater: the observers aren't in the same room; they're one floor up, looking down through large windows. I had to check it out.

This is a real farm, producing about eighty thousand pigs a year, so visitors get a genuine view of what life on many bio-secure farms is like. As I entered, a docent explained that the priorities of the farm are the comfort and safety of the pigs. The requirement of showering before entering the barns is mimicked in the entryway, before visitors, mostly young families, pass into long halls that lead to viewing rooms for each stage of pig raising. The walls are lined with posters that relate everything from the top states for pigs to the economic impact of pig raising (in Indiana, pig farms contribute more than three billion dollars a year to the state's economy) to facts about the pigs themselves.

In the farrowing barn, the statistics focus on what is happening on the floor below (when one litter starts to nurse, all the other sows are motivated to give birth; piglets weigh about three to four pounds at birth and seven pounds at one week; mortality rate among piglets is less than 5 percent, thanks to the anti-crush bars in the farrowing stalls; babies are weaned after twenty-one days). Swine technicians circulate in the farrowing barn, helping sows that are having difficulty with delivery, moving piglets in danger of being crushed or rescuing runts being attacked by their littermates, swabbing navels with iodine, and generally making sure all is well. Everything looks spotless, and a docent confirms this impression by relating that the floors are cleaned every three days, and once the piglets are weaned and moved out of the farrowing barn, the entire place is scrubbed and disinfected.

In the barns with young pigs, the stories about pig aggression come to life, with fights constantly breaking out around the room. Still, there is plenty of space, and most of the still-small, still-cute creatures focus on the two things they love best: eating and sleeping. In the sows' barn, the latest innovation in feeding can be seen. Sows will fight over food, and a timid sow can go hungry, so sows now have collars with electronic chips, and in the feeding area, a reader identifies the sow and makes sure it gets enough food—and that another sow doesn't steal it.

It is remarkable the degree to which every aspect of the pig's life has been considered. Everything is carefully controlled, from the temperature sows prefer when farrowing (67–71°F) to protecting piglets to providing clean air and water and a mix of food that will make all the pigs healthy and happy. While the viewing windows and focus on education are specific to Fair Oaks, this level of care is fairly typical of modern operations.

Processing Pigs

Once the pigs are grown, there is that transition from pig to meat that is inevitable. However, for most of those who are involved in this transition, there is the desire to make it as stress-free and painless for the animals as possible. But, in all cases other than whole-hog barbecue or pig roasts, that is just the beginning of processing.

Shorty's Locker is perched on a grassy rise overlooking the James River (or Jim River, as they call it locally), just outside of Mitchell, South Dakota. A sign announcing what is offered at Shorty's and the faint fragrance of hardwood smoke are the only indications of what occurs in the clean, white-roofed buildings on this rural road. Shorty's is a meat locker, a place where people can bring their livestock for slaughter, butchering, and however much processing is desired. Prices are posted on the door—thirty-two dollars to slaughter a hog, thirty-eight dollars if the hog weighs more than 325 pounds. Those who raise pigs understand the importance of having a professional handle this task, as it takes considerable skill and experience to kill a hog swiftly and humanely.

Also listed are prices for smoking, curing, and other processing, from wrapping for a customer's freezer to preparing for market. Inside, the locker's owner and operator, Nowell "Shorty" Hofer, guides a visitor around the front room. Hofer is a sturdy, animated man in his sixties, lively blue eyes watching for reactions as he points with pride to the impressive array: pepper bacon, ring bologna, chops, sausages, smoked hocks, pork roasts, hams, and a tremendous variety of bratwursts. The visitor buys enthusiastically.

When the visitor departs, Hofer invites me to join him in his office. Settling down behind his desk, he notes that when he started butchering hogs, it was because his family couldn't afford a butcher. "Mom and dad did it, and grandma and grandpa, back on the farm. In the 1950s, when I was a kid, there weren't a lot of meat lockers. We'd hang the meat in our own smokehouse all winter."

Hofer went into construction for many years, but it kept him away from home too much. When two local lockers closed, he decided that, since he knew how to process meat, opening a locker would solve his problem, giving him the opportunity to get rooted in the community. "Been here 18 years now," Hofer relates. "Built the place from the ground up. Started with a fifty-by-sixty-foot building, but now we're up to about 6,000 square feet." Today, he has six employees and handles three hundred hogs a year, all locally raised, plus beef cattle, buffalo, lamb, and goat.

"We have our own smokehouse here," Hofer continues. "We do beef, too, but when summer comes, people want pork. We make thirty-two types of brats and four kinds of smoked bologna. They all sell. For bigger events, folks want pork ribs or pork loin. If a customer wants something special, we can roast a whole hog for them. Roasting a whole hog is very celebratory. And the price is better when you buy a whole hog. There's less waste. We also make everything a person would need for the party, like salads, sweet potato and regular fries, chopped onion for your brats, everything. It helps ensure that quality is consistent, and it makes the customers happy."

Reminiscing about his childhood, Hofer notes that farming is hard work, but it taught him well—and had its advantages. "I've been in this building for 18 years, but I really started working with animals when I was 4 or 5 years old, helping Mom and Dad. Then I really did it on my own since the 1960s. We milked the cows first, and then we had breakfast at 7. We went back to work, and then dinner was at noon. Mom had a huge garden, and everything we ate was fresh. When I was young, we sold our cream to a company that made natural cheeses and butter. I still buy from them today."

Hofer handles processing both for retail and for individuals. The hogs all come from local farms, so Hofer can easily keep an eye on quality. "We work with the farmers. I'll tell them, 'If you bring something in that I wouldn't eat, don't bother unloading it.' I want everything we do to be top quality."

Like any other businessman, Hofer has to watch the trends, and in pork, demand has changed in recent years. "For a long time, people wanted lean, lean pork, but now a lot of folks are coming around to more fat again. The confinement operations are still raising leaner meat, because there's still demand for that, but there are also free-range farms now, where they raise breeds that have more fat. Both operations are good, they just meet different market requirements."

Hofer relates that there is a lot of German influence in the area, which helps explain the wide range of brats, as well as other German specialties. "One old German specialty is buckwheat grits patties. It's similar to scrapple in Iowa. You use the buckwheat to stretch your ground pork." A recipe is pulled out of a drawer and dictated. It makes ten pounds. "It's still popular. We not only sell it locally, but I ship it to someone in Illinois and another person in Texas. It's the sort of thing that, if you grew up with it, you miss it."[2]

Hofer likes those grits patties, but asked about his favorite food, he grins and says, "I'll take a pork chop any day. Bone in pork chop, with apples or cherries on the side."

Driving through the undulating green countryside in northern Wisconsin, I turn down a side road and pull up in front of a cluster of wooden buildings with a great view and the intoxicating perfume of smoke. This is the home of Nueske's Applewood Smoked Meats. The first Nueskes to arrive in the United States settled in Wittenberg, Wisconsin, in 1882. Wilhelm Nueske not only fed the family with glorious cured and smoked meats, he also taught his son John Nueske both the skills and recipes that had been brought from the old country. In 1933, John's son, Robert C. Nueske, realized that the meat he was eating at home was better than anything being sold commercially, so he launched "Bob's Fancy Meats," the company that would become Nueske's. As the meats became more popular, the whole family was enlisted in creating the products for which the company was quickly becoming renowned, even if only locally at that time.

Robert's son, Bob, would later say that he was "born in a cooler and raised in a smokehouse," which was a good thing, as he and his brother Jim were just in their twenties when their father died, leaving them in charge of the company. It was Bob and Jim who built the highly regarded but small company into a larger company with a national following and the kind of reputation where, if a restaurant buys it, they print it on the menu, so diners will know. In 2015, when Bob Nueske passed away, it was his daughter, Tanya, who stepped up to continue the Nueske family's vision. Fortunately, after years of working with her father, she too possessed the know-how and passion that had built the company. "I was fortunate that I got to spend twenty-four adult years working with my dad," Tanya Nueske shares. "It was wonderful for me, but it has also been a great thing for the employees to have seen that. They know there's continuity."

For many years, Nueske's meat, especially their bacon, was a semi-legendary food that would draw people from Chicago across the border into Wisconsin. Then, in the 1970s, the company decided to build a facility that was sufficiently large and modern to qualify for U.S. Department of Agriculture inspection, which would allow them to sell outside the state. Once the new facility was completed, the original sturdy stone smokehouse that was used during the company's first decades was moved to the new property, where it adds to the handsome, rural ambience of the new building, the exterior of which looks rather like a charmingly weathered old barn. That's on the outside. Inside, the retail area looks like a wood-beamed hunting lodge, but with wall-to-wall food, and behind that, there is a wonderfully modern facility designed to accommodate the family traditions that make Nueske's products so well loved.

Inside the impressively equipped production rooms, the most noticeable thing about Nueske's is that everyone seems happy. As Tanya Nueske moves from room to room, it becomes clear that this is very much like a family. She greets everyone by name, and they greet her. When she introduces Mike, the smoke master, she relates that she stood up at his wedding. Standing next to him, she enthusiastically describes his importance in the process. "He's like a European winemaker. There are thirty-four smokehouses, and they are in theory identical, but there are differences. It takes a lot of care and adjusting to create consistency, and Mike is able to do that—adjust the timing, the heat, the smoke, to make sure we always get the same quality."

Racks of cured pork bellies are rolled toward the smokers. "Some companies press the bellies, to make them more uniform for slicing and packaging," Nueske notes. "We don't press bellies. We feel using the natural shape and muscle structure preserves the integrity of the meat. It doesn't break down." Slabs of pork belly are being prepared in a room near the smokehouses, with slicing and packaging in the next room down. Following that are the rooms for sausages and hams.

While it is the twenty-four-hour family-recipe cure and twenty-four-hour smoking over sweet Wisconsin applewood that give Nueske's products their signature flavor, it also helps that they use really good pork. They have a private supplier who breeds hogs, Belgian Pietrain crosses, that have a good ratio of lean to fat.

Marketing Manager Megan Dorsch joins us in the production area and enthusiastically agrees with Nueske about the quality of the products. Dorsch adds, "People think of us as bacon, but we make so many other products. Our sausages are wonderful, and our smoked pork chops and smoked liver sausage. Our hams are popular locally for regional dishes, such as scalloped potatoes and ham. We also have products that we only offer here, because we want to keep this place a destination. For example, you can only get the jalapeño-cheddar brats here, and the ham salad. That's a local specialty. We have that 'log cabin' trailer out front in good weather, so people who are driving by can stop for a freshly grilled bratwurst and beans with bacon." Indeed, the setting is lovely, and it is hard to imagine being in the area and not stopping here.

"We're very true to who we were, who we have always been," Nueske states. "We have been encouraged to get really big, take shortcuts with the meat to produce more, move somewhere we could have a larger workforce. We don't want to do that. It's been nice to be able to grow and still stay in this community. This area is known for its fruit trees, and my grandfather loved this area. And he loved what the wood from the apple trees did for pork. It's hard work staying small, but we think it's worth it."

Kari Underly is a master butcher in Chicago. Her experience crosses generations, as well. She reminisces,

> One of my early memories was walking down into my *Busia's*—that's Polish for grandmother—cold basement and seeing her carefully preparing handmade Polish sausage. I was fascinated by the swift, gentle way she would turn what appeared to me like a pile of scrap into long, beautiful coils of sausage, which were then twisted into manageable links for the family. She would season the meat, delicately push it through the hand-cranked meat stuffer and into hog casings with perfect timing and coordination. When the sausage came through the end of the sausage horn, well I thought that was magic!
>
> I have always loved that memory; it was the first time I fell in love with the art of butchery. My Busia butchered on the farm during the Depression. My grandmother and grandfather on my mom's side were also butchers. And my

father worked his entire life as a butcher. I guess you could say that butchery is in my DNA.

Underly has seen changes, even in commercial butchery. She relates,

> Back in the 1970s and even 1980s, you could rely on a craft butcher to prepare an individual cut from a large piece of meat and package it right in front of you. Even supermarkets had their own meat cutters. The shift to shipping boxes of pre-cut, vacuum-packed meat has transformed the labor practices in our industry. There was no need for the personal service and the labor associated with it. Butchers were forced into jobs that require minimal skill or creativity. People left the craft. This shift has left an enormous void in the pool of quality meat cutters.

Underly observes that it is the increased interest in heritage breeds that is the real driving force behind the current snout-to-tail trend in restaurants, but it is also creating the increased need for people who know how to cut meat.

> Chefs are looking for older hogs, because the muscle conformation is tighter and holds less water compared to the younger commodity pigs. This is particularly important when curing pork. Restaurants are seeking out heritage breeds like the Red Wattle, Mulefoot, Ossabaw, Berkshire, and Mangalitsa. The Mangalitsa almost went extinct because of the huge amount of fat the hog produces, which was long unwanted, but today, chefs use it for curing.

Underly notes that the rise in the popularity of cured meats creates another area where greater training is needed. She emphasizes, "Curing meat can be dangerous and botulism is very real. Chefs and butchers alike need to understand the risks associated with producing cured and cooked meats. A food safety and HAACP [Hazardous Analysis Critical Control Points] plan is mandatory with charcuterie and fresh sausage programs."

Underly also notes that another trend that is creating challenges is the desire for local meat. "The challenge with getting the pig from the farm to the table actually happens before the butcher gets involved. Today, most of the slaughterhouses are in the center of the country, and we simply do not have enough local slaughterhouses to handle the demand for local meats."

So clearly, there is a growing need for well-trained meat handlers. The problem becomes how to train them. Underly explains,

> There are very few places that offer quality three-year apprenticeship programs, and there are no schools in the U.S. that train and certify butchers. However, there is no shortage of enthusiastic young people who want to learn the trade.

It is my mission to develop a school to train and certify butchers and meat cutters at all levels of the butchery trade. That's why I started Range, Inc., which combines consulting and marketing but also gives me a foundation for developing that school, where the next generation of butchers can be trained.

Specialty Pig

There are many ways of putting pig on the table: different cuts, different recipes, different traditions, both at home and in restaurants. Among those preparations enjoying a rapidly growing fan base are the related arts of charcuterie and salumi. Charcuterie has been mentioned as a hobby, but is more often a trade, or at least part of the arsenal of skills belonging to top chefs.

Both *charcuterie* and *salumi* are terms, one French and one Italian, for salted, cured, dried, or otherwise preserved meats, usually pork. Salumi includes salami, but it also includes prosciutto, lardo, mortadella, and the whole range of Italian-style preserved meats that are thinly sliced and served as appetizers. Charcuterie, which originally meant only "pork-butcher shop" in French, now commonly refers to the things one would buy in such a shop, such as sausages, smoked meats, pâtés, and confit. (In the United States, however, if one orders a charcuterie plate, one is likely to be served a blend of charcuterie and salumi.)

Brian Polcyn is definitely one of the stars in this realm, not just for his skill at curing meat, but also for the two do-it-yourself books he co-authored with Michael Ruhlman: *Charcuterie: The Craft of Salting, Smoking, and Curing* and *Salumi: The Craft of Italian Dry Curing*. These books are all that a budding *charcutier* or *salumist* needs to create the kind of taste treats that are increasingly offered on American menus. Polcyn, who is an award-winning chef, has now retired from running the two restaurants he once owned to focus on teaching a new generation what he knows—not just how to cook, but how to create the cured meats that one tends to associate with European vacations.

As Polcyn leads me through the gleaming, bustling teaching kitchens at Schoolcraft College in Livonia, Michigan, he checks on students and points out ongoing projects. Passing through the pastry kitchen he jokes, "It's not worth stopping here. It's just sugar, not meat." Not too surprisingly, in rooms where there is meat, there are also copies of Polcyn's books.

It was at the highly regarded Golden Mushroom restaurant that Polcyn gained his love of charcuterie.

I came up through the ranks, working in Detroit kitchens for seven years, but then I got to the Golden Mushroom in 1980. The chef there, Milos Cihelka,

> had amazingly high standards. He was the person who taught me how to think like a chef. They made all their own sausage and smoked meats, and studying with Chef Cihelka taught me how to do it and do it right. I loved the craft. It was so beautiful and so practical. I carried this love, and this practice, with me when I left, and I have enjoyed introducing charcuterie to a wider audience through my own restaurants.

Polcyn's goal is to introduce charcuterie and salumi to a modern American market—eating it, yes, but even making it. "Salami has been made the same way for 2,000 years, but because there are things like refrigeration and reliable sources of safe meat available today, we can now be more consistent and safer." He notes that charcuterie and salumi are practical because they use underutilized meats.

> We really do use everything except the oink. Pigs are perfect for charcuterie, not just because the flesh is so versatile, but also because the fat is so remarkable. But I use it all, the hearts, lungs, liver, intestines. My book *Charcuterie* shows how every part is used. My second book, *Salumi*, breaks the pig into eight parts: head, jowl, back fat, belly, shoulder, hams, neck bones, and scraps. I think the neck muscle is the best muscle on the animal. It gets more exercise, so there's more collagen. But everything is useful. The bones are used for broth. The scraps go into salami. We take these parts and make them taste better and last longer.
>
> A good butcher is a good cook. He can look at a muscle and tell you how it should be prepared. That's why we teach students how to butcher the pigs, not just prepare the meat.

Polcyn explains that there are actually multiple ways to butcher a hog. Every country has a different method. His book, *Salumi*, covers Italian and American styles, and each year, he represents the United States at "Pigstock," an event that offers intensive training in a variety of international butchering and curing techniques.

Talking about varied international approaches to food, Polcyn relates, "I grew up in a home with good food. I'm half Mexican and half Polish, and both of those cultures love good food. I still use my mom's pierogi recipe. In two of my restaurants, I've served Mexican food. But I really love focusing on seasonal food and European traditions, and that's why charcuterie became an important element of my restaurants."

Polcyn keeps track of Michigan farmers who are doing a good job of raising quality meat.

> Family farms are important, and I want to encourage those who are working at raising high-quality meat. We need to be aware that it's hard for anyone to make a living raising livestock, but it's harder for those raising heritage breeds. I'm working on building a USDA factory for charcuterie that will use heritage pigs, in an effort to make raising those pigs cost-effective for farmers, so farmers can raise them and still survive economically.

Polcyn's practicality goes beyond simply not wasting meat scraps. He also acknowledges the economics of pork.

> I love cooking with heritage pigs, but people need to remember that regular production pigs fill a need. They're wholesome and moderately priced. Not everyone can afford the really high-end stuff. There's also a place for mass-produced foods. We have to feed people. And even mass-produced food takes work. A hot dog is difficult to make. The heritage breeds taste good, but they put food prices out of range for average people. All my restaurants have been white tablecloth places. I focus on serving the best possible food, so I don't mind spending more for better meat. But there is a place for more modestly priced meat. It is more affordable, plus it's kind of a waste to use exceptional meat if you're going to sauce it heavily. But even then, you should raise pigs as well as possible. Don't shave nickels. Even production pigs can be raised well.

CHAPTER THIRTEEN

Cherished Pig

Traditions that Have Changed and Some Worth Keeping

Agriculture has changed more in the last 150 years than it did in the previous four thousand. In the Midwest, since the mid-1800s, we have gone from the ancient technology of a wooden plow pulled by two oxen and picking corn by hand to using combines that look like space stations. Granted, plows and seed planters had been modified and improved in previous centuries, but they were still pulled by animals until well into the 1900s.[1] In the mid-1800s, one farmer could feed three to five people. Today, each American farmer feeds on average 155 people.[2] And feeds them well. From ancient Greece through the Middle Ages and Renaissance, only a handful of aristocrats and monarchs ate as well as the average American does today, even on a budget.

The dramatic improvement in productivity helps us survive another dramatic change: the decline in the number of available farmers. In 1790, farmers made up approximately 90 percent of the work force. By 1840, that number was down to 69 percent, as more people became involved in industry and moved to cities. The terrible death toll of the Civil War further reduced the available workforce. The decline continued through the 1900s, with the Great Depression, the Dust Bowl, two World Wars, and economic fluctuations joining the attraction of urban opportunities to reduce the number of farmers. Today, less than 1 percent of the labor force is involved in farming.[3] However, one thing hasn't changed: the vast majority of farms in the United States are still family owned, though farms generally need to be much larger than in the past, since they are fewer in number.

The whole business of obtaining meat has been transformed, as well. From being a task handled by individuals on farms, processing meat has become an enterprise that is most often handled by professionals. How we raise livestock has changed, as well as what we raise. How we preserve food and how we cook it have, in some cases, been altered beyond recognition.

Many of the changes were improvements, but people have begun to recognize that something has been lost. Few people would be eager to go back to killing their own hogs and cooking in huge fireplaces, and it seems unlikely that anyone would want to return to letting pigs wander the city streets, but there are traditions that are resurfacing, some just as fond memories, but others as a reclaiming of worthwhile traditions. And, as always, new traditions are being created to replace the old.

Whatever else changes, raising pigs always involves people, individuals who reflect both the traditions and the advances, each one cherishing the stage of the process that belongs to him or her. We have moved on, leaving behind unworkable options, but we have also held on, trying to preserve the best of the past.

Different Paradigms

Humans have throughout history had specific goals when it comes to food. The first goal is not starving to death. The second is making sure others have enough. Next comes getting enough food to be healthy and energetic. After that, for many people, the goal is to eat well and then to eat better.

Through most of the 1800s, the focus in the United States was on not starving, especially among many of the immigrants who came from poorer countries. From the mid-1800s forward, the focus moved through having enough to eating well. Rationing during two World Wars plus the shortages of the Great Depression were bumps in that road, but today, while there are still people who struggle to get enough, the population as a whole has no fear of starvation.

Modern approaches to pig raising ensure that as many people as possible have a source of protein. More traditional approaches are intended to accommodate the desires of those who want a bit more luxury. Fortunately, because producers have options, so do consumers.

One of the options offered by modern advances is raising pigs indoors. Still, some choose to raise pigs *al fresco*. Each option has benefits as well as drawbacks, and there are debates as to which is better.

In Cass County, Michigan, a long, straight road passes through a gently undulating countryside where green and gold fields are dotted with silver grain bins and dense stands of trees. Just past a large, red truck shed, I turn onto a gravel drive where a sign reads "Jake's Country Meats." Nearby, the home of Nate and Lou Ann Robinson sits perched on a rise that overlooks the fields and groves of their sprawling farm. While it has evolved over the years, this farm has been in the Robinson family for six generations, and the seventh generation is in training.

The Robinsons raise heritage breed pigs, both purebred and crosses. Lou Ann Robinson relates,

> I grew up in a small town. I had worked on a farm as a kid, but never with livestock, so I had ideas but no experience. Then I met Nate. Coming into a livestock operation, I initially thought raising pigs indoors was the way to go. But over time, we both began to realize that outside was better for the animals, so we moved the pigs outdoors. We found that it was better for us, too, as we spend so much time outside. Then, in the 1990s, we began to realize we had something special.

Nate Robinson adds, "We had something special, but we weren't getting paid for special."

Lou Ann nods in agreement. "We made a lot of changes. We had to downsize and go greener, and then we had to go into higher-end markets."

As part of that effort, the Robinsons got AWA certified. Animal Welfare Approved (AWA) is a program that audits and certifies family farms that pasture-raise animals and that pursue exceptional levels of animal welfare in their farming practices. Farms that meet or exceed AWA standards are able to use the AWA label on their products.[4] "To keep our AWA certification," Nate notes, "we have to do a lot of paperwork to show where the babies come from. They not only have strict rules about how the animals are raised but also about how they're bred. Prohibiting genetically modified pigs is just the beginning of those rules."

Jumping into an old truck, Nate and I head out into the fields. "It's an open farm. Everything is outside," Nate explains.

> Sunlight is my best disease control. We vaccinate the pigs, but we don't use antibiotics. We use geothermal heat for energy, so we require no electricity other than for pumping water. We have eight miles of pipe to supply water everywhere it's needed. We seed the fields with grass, alfalfa, radish, turnips, those sorts of things. It gives the pigs a lot of variety. Of course, we're always reseeding, because pigs destroy fields. We also feed the pigs granola, because it sweetens the meat, and we add soy, because they need the protein.

Crossing the field, Nate points out differences between pigs. "Most of our pigs are Berkshires, with a sprinkling of Durocs, a few Hampshires, and some Yorkshires. We take meat to market based on what the market wants." He points out a few pigs with dark coats, one that is white with spots, and one that has lop ears and a ginger coat. "Different colors help us determine where we sell. Darker pigs have darker meat and more fat, so they'll be sold to restaurants in downtown Chicago. The white pigs have leaner meat, so they go to the suburbs. Darker meat and more fat tastes better, I think."

The fields are dotted with Quonset-like shelters, which offer pigs shade when they need it. Many of the sows have just farrowed, so all but a few are inside, nursing their piglets. Nate gets out of the truck to check each hut. "We never mix families," Nate relates. "They stay together. We have nine families right now." He picks up a small, black-and-white piglet. "Some people want the little pigs, but I hate to kill them when they're little. They're so cute. But I don't actually kill any of my pigs. I take them to a USDA site. The professionals can do it swiftly and humanely."

After checking all the huts, Nate jumps back in the truck and heads down a track through a stretch of forest. "Once the pigs are weaned, they will be taken to a different part of the farm, on the far side of these woods, where they're managed and trained to get by on their own. Plus, the distance between the areas limits the spread of disease."

The grassy field where the youngsters are trained has a few shelters, but also has a fair number of mud holes, which the pigs are enjoying. It seems a bit less civilized than the fields where the sows live, and in fact that is part of the idea. The "managing" they receive is essentially just keeping an eye on them while they learn how to be as close to wild as possible, though still comfortable with the presence of humans and the expectation of being fed.

The adjacent forest is not just for the pigs, and a fence runs through the middle. "Half is for pigs, and half is for wildlife," Nate explains. And the trees do more than shade the pigs. "When a pig goes to an elm tree and starts stripping and eating the bark, it means they're not feeling well. It can kill the tree, so we work hard at making sure all the pigs are well. But noticing signs like that can help you catch illness. Circling turkey buzzards are another signal. They alert us to a sick or dying animal. Nature is our monitoring system."

Walking past a field of lush, green grass, Nate relates, "Pigs do not like tall grass. They want to root for worms and grubs. So, my daughter suggested we use cattle to eat down the grass, and then move the cattle and bring in the pigs. We're about to bring our cows here to graze. Last year, we raised wheat here—soft wheat, like you use for Ritz crackers. By changing what we do with each field every year, we keep all the fields healthy."

Being outside, seasons and weather play a bigger part in caring for the pigs than would be the case if they were indoors. Nate explains,

> Pigs can tell when seasons are going to change, and even when the weather is going to change. Low pressure bothers them. It affects their attitudes. The sows are also more likely to farrow when a low-pressure system moves in. In the month of September, if you try to breed pigs, there is a higher incidence of stillbirths or miscarriages, because pigs can sense the approach of winter. Heat affects their attitudes, too. Pigs like to be cool, so we have sprinklers and misting systems, where they can cool off. If it gets really hot outside, we put cool water in a pool for the pigs.

Nate admits that there are drawbacks to raising pigs outdoors.

> Raccoons will take newborns. So will coyotes. We've seen 30 or 40 coyotes around the farm. Red-tailed hawks are an ongoing problem. They're here all year. Bald eagles are migratory, so they're only a problem in the spring. I think they bring their kids here to teach them how to hunt.
>
> Then there are the issues related to pig behavior. Pigs love to root. Michigan has a zero-tolerance policy for erosion, so we have to use nose rings with the pigs. The design has been changed, so the rings aren't as deep. They're called humane rings. The pigs can root with the tops of their snouts, which makes them happy but keeps them from doing the kind of damage that leads to erosion problems. It also reduces the risk of their uprooting a fence and escaping, though we still have to check the fences regularly. Michigan is working hard at not developing the kind of feral pig problem many other states have.

The Robinsons' farm is large for an all-natural farm. They usually have about fifteen hundred pigs, though their neighbors with indoor pigs have fifty thousand. It's not possible to raise really huge numbers on open land.

Back at the house, Nate and Lou Ann talk about the work involved. "Not a lot of people survive raising livestock," Nate explains. "A lot of the pig farms in this area have gone out of business in the last decade or two."

"We've worked hard," Lou Ann notes. "With heritage breeds, you sacrifice a lot. Berkshires have smaller litters. Outdoor pigs not only have smaller litters, they also take longer to wean. But we wouldn't want our kids to take this over if we couldn't make a go of it, earn a living. We've done that."

Nate smiles and adds, "They may be hard to raise, but they thrill my heart, even after forty-some years. We're cash poor, with our equity in the land. It's hard, but it's a choice. I think it's the right choice for us."

Confinement pigs, sometimes called production pigs or commodity pigs, are the pigs that produce the pork found in the average grocery store. The breeds being raised in confinement operations have been specifically bred to be more comfortable in enclosures. That is why breeders as far back as the 1700s began crossing their pigs with the more docile Chinese pigs, which had been confined for millennia and had developed traits that made them more suited to enclosures.[5] Also making them more suited to lives indoors is the fact that these pigs are leaner, which means they do not have the extra layers of fat to keep them warm in winter.[6] Confinement farms have improved steadily in the last few decades, but it is still important to remember that the pigs they are confining are not like wild pigs, or even like heritage pigs.

Pigs raised in confinement supply large quantities of safe, reliable, and affordable meat for a rapidly growing population. Also, fewer farmers must feed more people with less land.[7] There are some disadvantages, which will be discussed in chapter 14, but those who choose this approach feel the advantages outweigh the disadvantages.

Brad Greenway is a third-generation farmer and owner of Greenway Pork, a modern pig farm just outside Mitchell, South Dakota. His wife, Peggy, was a city girl, but after thirty years of marriage, she's as much a part of the farm, and as involved in the work, as her husband is.

Impressively modern pig barns stand amid green fields on the family-owned Greenway farm. The barns are, in fact, relatively new additions. Greenway relates, "We used to raise our pigs outdoors, but my wife and I wanted to do a better job of caring for the pigs. Indoors, we can control the climate, giving the pigs exactly the temperature they need. It's hard to keep pigs warm in the winter, and if they're not warm, it's hard to keep them healthy. Having the pigs indoors also makes it possible to give them a lot more individual attention."

It is clear that Greenway put a lot of work into finding the best options for his animals.

> That image people have of pigs in straw in outdoor pens seems ideal, but it's not ideal for large herds of pigs. Disease spreads faster on dirt lots, especially in cold, wet weather. We don't need nearly as much medicine now, with the pigs indoors. In addition to not getting cold, the pigs are also not exposed to animals that spread disease. Wild animals can't even get close. Pigs can catch bird flu, but birds can't get in here. Large fans make sure fresh air is circulating

all the time. Natural light comes in through windows and through the end curtains when the weather is warmer.

Food and water are monitored and delivered electronically, so pigs get plenty of fresh water and fresh feed. The nice thing about this system is that it means we can spend more time with the pigs, instead of measuring and moving food and water. We're with the pigs every day. We check each animal individually, to make sure it's healthy and getting everything it needs.

The Greenways utilize the latest research on how best to raise pigs. Pigs used to be moved to new facilities at each stage of life, but this has been shown to cause stress for the animals. The barns here are wean-to-finish—from the time the pigs are weaned until they're ready for market—so the pigs stay in the same facility. Plus, they aren't being mixed in with new pigs all the time, which causes even more stress.

"We keep the pigs together by size," Greenway explains. "A smaller pig would become a target if kept with larger animals, so we put any runts or timid pigs in with pigs their size. We have large pens, so pigs can move around and socialize with others. And they always have access to food and water."

One of the things the Greenways are striving for is sustainability, and that is something this system offers them.

This is better for the environment than the way we used to do it. It took a lot more food to raise the pigs outdoors, because it takes a lot of energy to stay warm. Also, with the pigs indoors, we can collect all the manure and use it on our crops. We grow 800 acres of corn and another 800 acres of soybeans, and the manure from the pigs can fertilize about two-thirds of that land. Then we can turn around and feed the crops to the pigs. There's nothing to get rid of, and the manure is immensely good for the corn. Plus, it saves us thousands of dollars—savings that we can pass along to consumers.

Handling manure involves a lot of planning, as well as expertise. "The Department of Environment and Natural Resources controls how much manure can be used," Greenway explains.

Prior to applying manure to our crops, we have to test the soil for nutrient need and the manure for nutrient value, so we know just how much manure can be applied. We hire a firm to collect and apply the manure. The manure is actually injected or mixed into the topsoil, which reduces the smell and reduces runoff. Still, there is some smell, so we call our neighbors when fertilizing season begins, to let them know they might notice it. I go around and talk to them and answer questions, so they know what's happening. We want to be good farmers, but we also want to be good neighbors.

Greenway Pork sells about ten thousand pigs a year, so it's feeding a lot of people. "Farms are bigger," Greenway notes, "but they're also a lot more efficient. A farmer in 1950 fed 27 people. Today, that same farmer feeds 155 people."

Greenway has done everything he can to protect the pigs and the environment, and to make sure people have access to food that is safe, affordable, and readily available. "If I weren't proud of what I'm doing, I'd stop."

The only possible answer to "which is better?" is "it depends." There are so many variables. As with most things in the real world, there are no easy answers, and there are problems and advantages with all approaches. But it is reassuring to know that farmers are very serious about doing the best job possible to reach the goals they've chosen and that people are constantly searching for ways of improving how animals are raised. But it is also good to be realistic, and to understand that certain things are impossible. We can't feed everyone with pasture-raised heritage pigs.

Processing Traditions

There was a time not so long ago when all pork processing was essentially a community affair. The community might be just the family or the whole neighborhood, but killing, cleaning, and processing a pig was a great deal of work, and it had to be done quickly, so as many hands as possible were needed.[8] That tradition still exists in some places, but it is now far from common. However, there are still a lot of people who grew up with that tradition, and they have fond memories of the outcome, if not always of the process.

Alice Mae White, mother of four and grandmother of eleven, grew up on a farm in Arkansas. Then, like many African Americans from the rural South, she moved north, settling in Detroit, Michigan, because wages were better. However, she brought with her a love of cooking soul food, especially greens, and vivid memories of the annual hog killing on the family farm. Her mother had died when she was a child, and Alice and her three siblings were raised by their father, whom she reveres. Alice settles happily into recounting those days, her strong frame straightening and smile widening as she describes what she still considers to be a vastly superior way to obtain one's pork, a deep, happy laugh punctuating her stories.

> We raised a few pigs, but my daddy also worked for a man who had hogs. Every year, as a Christmas bonus, my daddy's boss would give him a hog, and my daddy and his twin brother would split it. Daddy and Uncle Elbert would split the work and split the meat. I'd ask, "Daddy, you goin' to kill him," and Daddy

Photo 13.1. The snow drifting down in this photo is a reminder that for most of history, people waited until early winter to butcher hogs. In post–World War II Germany, Ella Bernhardt, aged four, examines the pig being given as payment to her mother and aunt for work done on the farm. Paying debts with food has also been common throughout history. Soon after this, Ella and her family moved to the Midwest. Photo courtesy of Ella Bernhardt.

would say, "We gotta eat something." We'd feed the pigs corn before killing them, because it sweetens the meat. Then Daddy and Uncle Elbert would get working on that hog. They'd cut it up, cover it with salt and some purple stuff,[9] and then hang it in the smokehouse.

We'd eat some of the pork fresh. I remember those pork chops; they were real good. But most of the meat got smoked. In the morning, someone would just go out to the smokehouse and cut meat off something hanging in there, and we'd have it for breakfast. You can't find meat that good anymore.

Granny did a lot of the work, too. She'd be out back with a cast iron kettle, and she'd make cracklin', and we'd sprinkle salt on it and eat the cracklin', and we'd use some of it to make cracklin' bread. I loved cracklin' bread. Granny would also hook up a water hose to the intestines and clean them out, and she'd scrub them, and then we'd put them in a bucket to soak, and we'd have chitlins for Christmas and New Year's.

Granny cooked pigs' feet, too. And the snout. She'd fry it till it was crispy. We'd use "everything but the oink," that's what Daddy used to say. We got lard from the hog, too. Between the pigs and what we grew in our garden, we had all the food we needed.

> We had a great big, old pig, as big as a table. Daddy said she was too big to be good meat. Just good for having babies—and she had lots of babies. I remember holding them in my arms. There were pigs in the yard. When the pigs got turned out into the field, we never had to worry about snakes, because the pigs ate the snakes. We had a tire swing hanging from a tree, and we'd slide down the hills in cardboard boxes. You always had something to do. I was a country girl living in the country. We had fun. Kids don't have fun anymore. And the food was always good and family was always around. It was a good life.

When towns began to grow, the job of processing pigs was often turned over to a local butcher. The traditions surrounding home butchering might have been missed, but less mess and the greater ease of having the work handled by a professional made this a task that was readily given up by most. Many of the little, old village butcher shops are gone, but some remain.

In 1976, the National Park Service named Drier's Meat Market a National Historic Site. At that time, the butcher shop in Three Oaks, Michigan, had just celebrated its one hundredth birthday. The narrow, deep building is little changed from a century ago. A steel railing with the outline of a pig has been added, to help customers get up the two steps into the store, but otherwise, the look is the same as it is in the aging photos inside: big front windows, white-painted wood, a sign in the window advertising homemade liver sausage. Inside, wood-plank walls are painted white and gray. The paint is clearly fresh, but the color scheme matches that seen in the abundant old photos. The floor is dark wood and scattered with sawdust. The wooden counters of the 1800s have been replaced by refrigerator cases, but Old World–style sausages still fill those cases. And the person behind the counter is still a Drier.

Carolyn Drier, the current owner and operator, clearly enjoys the market's historic status. Speaking enthusiastically about earlier generations, she points out old butchering tools and an antique meat rack, newspaper articles, awards, photos, letters, and memorabilia hanging on the walls. Carolyn is the third generation of Driers to run this shop. The building actually predates the Civil War but was converted from a wagon shop to a butcher shop in 1875. It was known then as the Union Meat Market, a name that still appears above the door. Carolyn's grandfather, Edward, Sr., started working there in 1899, at age ten. When the owner decided in 1913 that he'd rather be a banker, he sold the shop to Edward. From Edward Sr., the shop passed to Edward Jr., Carolyn's father, and then to Carolyn. They can all be seen in handsomely framed photos that illustrate the work that is still carried on here: cutting meat, stuffing sausage, stoking fires for the smoker.

Photo 13.2. Drier's Meat Market in Three Oaks, Michigan, is a National Historic Site. Photo by Cynthia Clampitt.

Talking about her father, Carolyn says,

> My dad appreciated the importance of advertising, and he was good at it. He also realized that, once the highway was built that bypassed Three Oaks, we'd have to do something to make it worthwhile for people to make a detour. He said that people weren't going to go out of their way for four pork chops. He was right. That's when we stopped selling fresh pork. We just do specialty meats now. We already had great smoked meats, and we just focused on those. We've got what we call our liverbest—instead of liverwurst, because we think it's the best liver sausage you'll ever have. It's more like a pâté, really. Hot dogs and bratwurst are popular in the summer. And people want our hams for the holidays.

That is one thing that has not changed—the seasonality of the business.

"Sausage stuffing and smoking are still handled the same way," Carolyn notes. "We even have the same smoker we were using a hundred years ago." But some things have changed. "When my father started here, we had to slaughter our own pigs. Now, we receive just the meat we want, not live pigs, and not whole pigs. We still have to break down large sections, such as slabs of pork belly or whole hams, but we don't have to kill the animals." A change that Carolyn introduced was putting the shop on the Internet. Driers.com makes their thick-sliced bacon, liver sausage, double-smoked ham, souse, and Polish sausage available to locals who have moved away or fans who have just passed through and gotten hooked on the secret family recipes.

Carolyn's oldest daughter, Julie, and nephew, David Wooley, both great-grandchildren of Edward, Sr., decided the legacy needed to continue, so they came on board to help Carolyn. "Owning the shop is hard work," Carolyn admits, "but it's really nice to be here. We became a historic site in 1976 because the building was a butcher shop for 100 years. In 2013, it had been in our family for 100 years. You don't see that very often. It's not always easy behind the counter, but when people come in and tell us their grandparents shopped here, and they thank us for still being here, that makes it worthwhile."

Recaptured Tradition

Though convenience and economy were (and still are) in demand, some folks realized that costly, time-consuming, labor-intensive traditions offered different benefits—primarily taste. Among those who have in recent years embraced an older processing paradigm is Herb Eckhouse, who, with his wife Kathy, founded La Quercia Handcrafted Cured Meats.

La Quercia (pronounced la KWAIR-cha) is Italian for "oak tree," which is triply appropriate because the oak is the state tree of Iowa, where the company is located; it is a symbol of Parma, famed source of some of Italy's best prosciutto; and it is also the source of one of the oldest traditional foods for pigs: acorns. (Though, interestingly, the pigs in Parma are not fed acorns. They're fed the whey from making Parma's even more widely famous product: Parmesan cheese.[10] It is Spain's illustrious *jamón ibérico* that is made from pigs raised on acorns.)

For Eckhouse, appreciation for the abundant good pork in Iowa, combined with a love for foods discovered while living in Italy, led to the dream of starting La Quercia. Eckhouse feels that "we should have the attitude they have in Italy, of using the ingredients we have here, to highlight what is great in the region."

When Pioneer Hi-Bred, the seed company for which he worked, asked Eckhouse to head up their Italian branch, he had no vision of being seduced by the food of Italy. "In Italy, they make a cult of mothers, children, beauty, and food. We lived in Parma for three and a half years, and it was there that I learned how to eat. We had our two kids with us, and we certainly wanted to explore the art and history, but I also learned in time that people in Italy get up every day expecting to eat something special."

The seed of what would become Eckhouse's dream was planted while he was living in Parma, a city with a remarkable culinary heritage.

> The Italian attitude about food, and the taste of their prosciutto, got me thinking—and researching. I did a five-month study of trends, and I realized there had been substantial changes in the culinary focus of the United States. American wines and cheeses were becoming world class, and so I figured there was no good reason it couldn't happen with cured meats. My decision seemed affirmed over lunch with a friend from Venice. We were eating prosciutto, and it was so good, we ordered a second plate. My friend held up a sliver of the meat and said, "You know, if you could make something this good, you could make a lot of people happy."

That was the year 2000. Eckhouse would spend the next five years exploring both what people wanted and how to create it. Eckhouse began importing different kinds of prosciutto, to find out what flavor profiles were most appealing to U.S. consumers. Then he began learning how to cure pork. It takes up to a year to transform a pig's leg into prosciutto, so this was not an overnight experiment. The first experimental hams hung in a guest bedroom to age, but soon the Eckhouses built a state-of-the-art facility and made the leap into full production. Their first prosciutto shipped in 2005.

"My wife Kathy is not the risk-taker in the family," Eckhouse relates. "However, she's the one with the greatest culinary knowledge, so I couldn't have done it without her. She was cautious. It was my dream, not hers. She was supportive but not ready to commit at first. Then, after a few months, I told her I needed her to come on board, and she did. We're full partners in this enterprise."

Eckhouse emphasizes that he's not trying to compete with European styles. "We just want to make the best possible prosciutto in Iowa—as good as European prosciutto, but carrying the American banner. We can showcase how great the produce is that we have here. We wanted from the start to demonstrate that, to show the quality that America is capable of producing."

And while Eckhouse may not be competing with Europe, there are parallels that help guide his decisions. Eckhouse explains,

> Iowa used to be an oak savannah. It was our goal from the start to bring acorns and pigs together. We found a grower, and we brought the acorns to the pigs. Now we've found someone with a large wooded area, so we can take the pigs to the acorns. There is no way we could do this with all the pigs we use—there just aren't enough acorns available. In fact, only about ½ to 1 percent of our pigs are acorn fed. But the hams from those pigs are remarkable.

With acorn-fed pigs, Eckhouse buys the whole pig, but only cures the rear legs. The rest of the meat is sold fresh to local outlets. For the other pork they need, they buy individual parts from their suppliers. While they make a range of cured meats, La Quercia is best known for its prosciutto, and they use a lot more hams than bellies.

> We are always working at developing new sources, to help us develop new flavor profiles. We collaborate with cheese makers, who feed whey to their pigs. One farmer we know uses his pigs as his primary means of disposing of whey. Then there are different breeds of pigs. Berkshire pork offers a deep, red color and lush mouth feel. The Tamworth has meat that is very sweet when it ages. We're now getting a Berkshire cross that has a distinctive taste. Only our Prosciutto Americano is not breed specific.

La Quercia was not the first company in the United States to make Italian-style, dry-cured meats, but it was the first company to use only organically raised pigs. "We don't buy confinement pigs, Eckhouse continues, "and we don't buy pigs that have been given antibiotics. We want our meat to be as close as possible to what it was when the traditions were started in Europe—and maybe even better."

Prosciutto actually has much in common with what we in the United States call country ham. It's heavily salted and aged for years. Some American restaurateurs are now serving country ham and calling it American prosciutto. So, to a certain extent, we've made prosciutto here since the early days of colonization. We just didn't eat it like prosciutto. But it is just one of the traditional ways of preserving pork that took root here.

For roughly half a century, U.S. Department of Agriculture efforts to prevent the spread of diseases in pig populations included restrictions on meat products imported from countries where those diseases were an issue, including several regions in Italy. So up until 2013, when improving health practices led to some specialties being approved for importation, many of the Italian-style meats we enjoyed in the United States had to be made here.[11] One venerable Midwestern company that has long filled that gap is Volpi, a family-owned operation that brought generations of Italian tradition to St. Louis in 1902 and are still supplying Italian specialties more than a century later. However, now that some importation bans are being lifted, we hope that Americans don't forget those who have fed us all these years.

Whichever choice one makes—imports, venerable American offerings, or new but traditional high-end specialties like those from La Quercia—there are more options today than ever before. So, while the tradition of cured meats goes on, difficulty in getting them has ended.

Teaching Traditions

Among the truly important traditions to pass along is farming—or at least an appreciation of farming—and a number of organizations are working to accomplish this. Aside from agricultural colleges, there are local programs for a range of ages that contribute to this endeavor.

Students interested in pork production can become Pork Ambassadors, a program run in many states by the professional pork organizations. Becoming a Pork Ambassador offers college students scholarships and opportunities to be involved in the industry, making connections and learning what the roles are. In many areas, even younger students can learn about farming and raising livestock, thanks to organizations such as 4-H and FFA.

4-H, which is part of the university extension system and U.S. Department of Agriculture, is the country's largest youth development organization. It has a presence in every county and parish in the United States, offering clubs, camps, and afterschool programs in partnership with 110 universities. While originally created to reach rural areas, 4-H is now also active in both suburban and urban communities, promoting the development of the four Hs—head, heart, hands, and health—to six million youth across the nation.[12]

FFA used to stand for Future Farmers of America, but in 1988, the organization became National FFA, in acknowledgment of the fact that our food system is no longer just farmers, but has a range of careers, including technology, medicine, marketing, research, and more. Thus, the goal has changed from simply creating more farmers to supporting and encouraging all students who might be "interested in agriculture and leadership."[13]

Tyler Stenjem of Cambridge, Wisconsin, got hooked on FFA watching his two older brothers raise pigs as part of the FFA Swine Project. "Our swine project is open to any student in the FFA, from 7th grade to a year out of high school, who wants to raise and train a pig to show and sell at the Jefferson County Fair. Growing up on a family dairy farm, I had a fondness for animals, and I just took to pigs. I started showing pigs when I was in middle school, and it had a huge impact on my life."

Stenjem relates that students generally buy pigs that are about two months old from local breeders. "But my senior year of high school, some friends and I decided we wanted to actually breed our own pigs to show. That turned out to be a pretty harrowing experience, especially since we didn't have any facilities and tried to do it at home, which was rough during the winter, when we couldn't leave them outside. The house got pretty smelly. It was, however, an eye-opening experience."

The FFA program in Cambridge began to lag. Because he believed so strongly in the opportunities it offered to develop a wide range of useful skills, as well as the lessons about responsibility and pride in a job well done, when Stenjem graduated, he volunteered as the FFA Swine Leader.

> I've been the project leader going on 8 years, and it's really rewarding seeing kids in the program grow and mature. It's fun to look back and see what a kid was like when he or she first started and see where they are when they complete the project. Most of these kids will most likely never raise another pig, but at least they get the experience of what it takes to raise a quality product for the dinner plate, as well as an idea of the markets.

The price of show pigs began to go up, and it was getting hard for some of the kids to afford a pig for the project. So, when a friend called Stenjem and suggested they start breeding pigs, he saw that as a possible solution to the problem.

> I realized it would be a great opportunity. Not only for us, but also for the kids, who could then experience everything: breeding, gestation, and farrowing, in addition to raising the pig. Our FFA chapter is lucky enough to have the Severson Learning Center, which includes a 54-acre farm, so if kids don't have a place to keep pigs, they can keep them there. We currently have three sows we are breeding, a Berkshire, a Hampshire, and a Yorkshire. We also have a registered Duroc boar. The majority of the breeding we are doing is crosses among these four.

Stenjem recommends the FFA for anyone who wants to grow personally, noting that there is more than agriculture to the organization. Still, he thinks raising pigs is worth trying, at least once. "Pigs can teach you a lot about yourself, as well as offering an opportunity to develop skills you will use the rest of your life. It will also remind you, next time you sit down to dinner, just how much work goes into putting meat on your plate."

Fourteen-year-old Brooke Parish is one of the students involved in the program Stenjem leads. Brooke's mom, Raquel, says that even though Brooke is aiming for a career in marine biology, she had a blast caring for her pigs and participating in the county fair.

> Brooke learned a lot about patience with animals that she never had before. Pigs tend to do what they want and are very difficult to train, so Brooke had to work hard. She had to keep the pigs clean, and she had to feed and water them twice a day. She also had to build rapport with her pigs, and train them to show well in the ring. Brooke also had the opportunity to build strong relationships with other students.

Asked what the fate of the show pigs are, Raquel Parish notes that some are sold for breeding, but most are sold for meat. "After exposure to other animals at the fair, it's too dangerous to reintroduce the pig into a herd, because of the disease threat. In Brooke's case, she sold one pig for meat, but another went to Tyler, to breed. In exchange, Brooke will get two piglets when Tyler's sow farrows. Now, she plans on showing pigs until she graduates."

Ken Ropp, a dairy farmer in Normal, Illinois, had a different outcome when his daughter decided she wanted to raise a show pig. It was he who became hooked on hogs. His daughter thought that the whey from Ropp's cheese-making operation would be great for feeding a pig. She was correct. Today, Ken raises heritage pigs for local restaurants, and he no longer needs to look for ways to dispose of his whey. His small herd of fifteen pigs includes Chesters, Yorkshires, Old Spots, and Durocs. He tried Berkshires, but they wouldn't fatten on whey. Chesters are his favorite purebreds. Today, he says

Photo 13.3. Brooke Parish guides her pig Duke into the show ring at the Jefferson County Fair in Wisconsin. Photo courtesy of Raquel Parish.

demand outstrips supply, so he's thinking of adding a few more sheds so he can raise a few more groups of pigs. So clearly, having kids involved with pigs can have a far-reaching impact.

Cherishing the Past

While we cannot keep all traditions alive, there is a degree to which we can still visit them. The Midwest is home to some remarkable museums and living history venues, where one can see what things were like during the early days of the region. And because pigs were so important during our early history, they are in evidence in most places, and actually on hand in some. These are just a few examples.

At the National Road Museum in Norwich, Ohio, an extensive diorama shows both the history and extent of the National Road. In that diorama, on a muddy stretch just being cleared to begin the road, a herd of pigs approaches an early tollhouse. At Greenfield Village in Dearborn, Michigan, an 1828 tollhouse has been restored. The sign on its side shows tolls for using the bridge across the Merrimack River: eight cents each for hogs.

The Living History Farms in Des Moines, Iowa, take visitors through many layers of Midwestern history, from the 1750s to 1900. In the 1850s farm, which is the classic pioneer farm with split-rail fences and log cabin, dark-coated heritage pigs snuffle about in a split-rail pen, rooting in the dirt and tearing into the vegetables they've been given.

In Eagle, Wisconsin, the state's history from 1840 to 1915 has been preserved at Old World Wisconsin (OWW). More than sixty historic buildings have been brought here from all over the state, and interpreters help recre-

Photo 13.4. A diorama at the National Road Museum hints at the importance of the road for moving livestock. While only a few pigs are shown here, thousands of pigs moved over the road. Photo by Cynthia Clampitt. Used with permission.

ate that sweep of time for visitors. Dirk Hildebrandt is the Historic Farmer in charge of all agricultural programs for OWW, and he shares about the heritage pigs they have on site. “Currently, we have Ossabaw Island hogs and Mulefoots, all breeding pairs. Old World Wisconsin has always striven to have the most typical animal breeds based on period and ethnicity for the farms here. In the past, we’ve had Chester Whites, Tamworths, and Berkshires—and sometimes the accidental crosses that can arise.”

Hildebrandt relates that having pigs at OWW was important, because that is true to the period.

> The Wisconsin frontier began to be settled by the 1830s, and pigs were ubiquitous. Everybody ate pork. The different European ethnic groups that settled the state had eaten pork in the old country, and they continued to do so in their new home. The variations in cuisine that did exist were most evident in sausage. All ethnic groups made sausage, but everyone, even down to the family level, used their own recipes and different blends of spices. The Finns liked their cardamom. The Germans used a lot of brown sugar. Nutmeg is found in many Norwegian recipes.

Near one of the vintage barns, a shaggy-coated sow seeks the comfort of a mud puddle while her piglets explore the rustic enclosure. All pigs at OWW are in era-appropriate barns and pens. Hildebrandt explains,

> Turning pigs loose in the woods to fend for themselves was not common here. The frontier could be a dangerous place, and pigs were too valuable to lose. It was much more the norm to keep them in a sty close to home. Besides, pigs that run about won’t gain weight nearly as fast. People would keep as many pigs as they needed, either for meat or for sale. And for lard. Can’t forget lard. Early Wisconsin settlers didn’t rely on pigs exclusively, but pigs are fairly reliable, and any sort of reliability on a frontier was welcome.

CHAPTER FOURTEEN

Problems and Promises

Issues, Concerns, Discussions, and Hopes

The history of pigs is replete with tales ranging from inconvenience to disaster. While problems have been addressed over time, and some have been solved, problems continue to arise. Growing human populations mean less land for agriculture but greater demand for food. Improved incomes worldwide have led to increased demand for meat. As more pigs are raised, more people get involved in facing the challenges.

The Problem of Poo

As in the late Middle Ages, so too now, much of the discussion about having pigs revolves around how to dispose of what pigs dispose of. Fortunately, in most places, pigs are no longer on city streets—but there are a lot more pigs. There are, in fact, roughly sixty-six million pigs and hogs on U.S. farms. That means a lot of waste. Fortunately, most of the livestock community (farmers, scientists, universities, the U.S. Department of Agriculture [USDA]) has been working on this for decades, and the technologies now either in place or being implemented are making huge improvements.

Veterinary clinician Dr. Steve Henry, who specializes in swine and population medicine, has seen a lot of changes since he began practicing in the 1960s. He is optimistic about the current innovations in handling pig waste and particularly appreciates changes to improved systems for utilizing that waste.

Much of the history of agriculture has revolved around supplying crops with the nutrients they need, especially nitrogen and phosphorus. In the United

States, these have traditionally been mined or imported, because it's not possible to gather manure from free-ranging animals. During World War II, a method of creating ammonia, the most common way of delivering nitrogen, was developed as part of the war effort (because it's also used in explosives), which made it more readily available, but it still had to be piped or transported to farms. Modern pig farms and new technologies are changing this.

Henry notes,

> We produce meat, but we also produce manure and urine—both of which are very rich in nutrients that benefit crops. Back when I started, pigs were raised outdoors, and all that nutrient wealth got wasted. Worse than that, when it rained, that all ran into streams and rivers. The phosphorus pollution from the pig waste caused algae blooms that ruined the water. We now add an enzyme to pig feed, so a pig can absorb all the phosphorus it needs from the corn it eats, so no additional phosphorus needs to be added to feed. This dramatically reduces the amount of phosphorus in cases where there is still runoff. Moving pigs indoors helped even more in reducing the problem of pollution.
>
> Lagoons for waste were the next step in the evolution of dealing with the waste. However, that wasn't perfect, either, because sunlight causes ammonia to be released into the air, and the phosphorus sinks. So now, instead, we have huge septic systems that gather and hold all the waste until it's needed. Each tank produces the equivalent of $20,000–$80,000 in fertilizer. And there's no runoff. Plus, there is now no need to pump ammonia across the country or mine phosphorus. It lowers our carbon footprint tremendously.

Septic systems helped solve another problem presented by lagoons. In areas where there is a lot of rain, lagoons can be overwhelmed. This is a particularly big problem in North Carolina, where summer hurricane season regularly brings torrential rain, leading to leaking or collapsing lagoons.

"We faced many of the same issues with raising pigs sixty years ago, when we had small, diversified farms," notes Dr. Paul Walker, an environmental management consultant, recently retired investigator for the Livestock and Urban Waste Research Team, and professor emeritus at Illinois State University. "But we didn't have the concentration of animals or waste back then. Manure is a nutrient until there's too much, and then it's a pollutant."

While septic tanks solve some problems, Walker points out that there are issues they don't resolve. One is nutrient overload. In some places, there isn't enough land within a reasonable distance of a farm to use all the fertilizer. Another problem is odor. Pig waste really stinks, and while it is often injected into the soil or mixed into topsoil, to minimize the odor, it's still pretty unpleasant. Pig farms are on average 150 miles away from large population areas, but there are still plenty of people living in the areas where pigs are raised—and as urban areas continue to spread outwards, more folks might find themselves downwind from that smell.

Walker relates,

> People have been thinking through things more in recent years, and they're getting a much better handle on the issue of nutrient overload. But odor is still a big issue. Separating liquids and solids helps, as does composting. Methane digestion can turn waste into an energy source, and drying it might create a fuel source. The industry is trying to tackle all these issues. But these are costly alternatives, and most farmers can't afford them. That's one of the problems: cost. Either the farmer goes out of business or the price of food goes up. But eventually, we'll have no choice.

Costly or not, there are those who are charging ahead with these new alternatives. Research stretches back into the 1990s, and by 2000, several of the largest American swine producers were testing liquid-solid separation, nitrogen management, and systems that would remove pathogens and heavy metals from pig waste.[1] In 2013, Smithfield Foods reached an agreement with St. Louis–based Roeslein Alternative Energy to create a system in their northwest Missouri properties to capture and purify methane produced by their pigs. The Smithfield facility is one of the largest hog-feeding facilities in the United States, with nearly two million pigs. The first phase of the project was finished in August 2016. The project will be a source of renewable energy, producing a huge amount of natural gas—the equivalent of about seventeen million gallons of diesel fuel.[2]

Farmers are updating at different speeds, but everyone understands the importance of making changes—because the people who are closest to the problems are the pork farmers themselves.[3]

Disease: Theirs and Ours

For many years, in the United States, the most widely known ailment carried and transmitted by pigs was trichinosis. While one should still be careful with pork from wild boars or foraging pigs, among farm-raised pigs, this problem has been eliminated. Trichinosis became an issue in the early 1900s because of the practice of using pigs to "process" urban waste. A study published in 1920 revealed that about one-third of American cities with populations over one hundred thousand used pigs for garbage disposal. This was not really a new practice. Throughout history, pigs have been used as urban cleaning crews. However, it was more organized by the early

1900s—and the benefits were more than just less garbage. Cities produced tons of garbage, and every fifty pounds of garbage could produce one pound of marketable pork. Garbage-fed pigs became an urban income stream. Unfortunately, a lot of the garbage came from restaurants. While meat had long been fed to pigs in feedlots and on farms, it was always cooked meat. But waste from restaurants included raw meat scraps. It was an efficient way to handle garbage, and selling the pork brought in money, so cities were not eager to give up the practice. The public was simply told to cook their meat well enough to kill the parasites.[4]

Trichinosis was not the only disease transmitted by consuming infected meat, but the other diseases, including foot-and-mouth disease and African swine fever, only kill livestock, not humans. However, a virus that struck pigs in the early 1950s raised concerns about the possibility of an epidemic. In 1955, the USDA decided "eradication is vitally important" and passed a law that all garbage had to be cooked, to kill anything that might be in the raw meat.[5]

Though garbage-fed pigs represented only about 2 percent of the pork being sold in the United States at the time, they caused the greatest proportion of health problems. Eventually, not only did cities disallow the selling of garbage-fed pork, even commercial farmers were no longer permitted to serve slops to their swine.[6]

New feeds were developed. Soy began replacing the protein originally obtained from meat, but that wasn't enough. The "magic ingredient" that made the difference was vitamin B_{12}. Vitamins weren't discovered until 1912, but B_{12}, which is only found in meat or milk, was not isolated until 1948 and wasn't synthesized until the 1950s. B_{12} was what was needed to keep pigs healthy. (All omnivores, in fact, need this vitamin.) After the 1950s, it was possible for the first time in history to raise pigs without feeding them meat.[7]

Today, commercially raised animals receive carefully formulated feed designed to protect both pigs and consumers. South Dakota pig farmer Brad Greenway notes, "Pork is so safe now, the government has dropped the requirements for internal temperature for cooked pork from 160 degrees to 145 degrees."

Another reason we worry about disease in pigs is that, because we rely so heavily on them for food, if they got sick and died, it could potentially be catastrophic. (This possibility was underscored in 2015, when a bird flu epidemic led to the death of millions of chickens and turkeys.)[8]

Among diseases spread by pigs, influenza has historically been the greatest problem. There is some debate as to whether influenza originated in pigs or pigs got influenza from humans very early on and spread it. Either way, the introduction of influenza into human populations is associated with the beginning of domestication of animals. While pigs are not the only other animals besides humans to get flu, they are remarkable reservoirs for flu viruses, particularly the avian H1N1 and human H3N2 influenza viruses. Because pigs can play host to avian and human viruses, as well as viruses for swine flu, it is theorized by some virologists that influenza viruses hosted by pigs can combine and mutate, forming new viruses.

Epidemics and pandemics of influenza dot history, with the most recent pandemic occurring in 1968–1969. One of the deadliest plagues of all time was an influenza outbreak in 1918, with a death toll estimated at between fifty and one hundred million people worldwide. The influenza virus was actually discovered in 1918, not by studying infected humans, but by studying pigs, which seemed to be suffering from the same disease that was killing so many people. There are pigs today that still carry virus strains that are direct descendants of the 1918 influenza virus, though the pigs have become immune to that specific form of the virus.

Vaccines have been developed, but the best way to protect humans and pigs is to keep pigs away from anything that might spread infection, including birds, other animals, and even many humans. In most of today's big confinement operations, workers scrub up and change into sterile clothes when entering a pig enclosure. Everything is kept clean, to keep the pigs safe. Even the air is filtered. Because one stray bird or feral hog with a new influenza strain could introduce a virus that would spread through the whole herd.[9]

Previously mentioned was the porcine epidemic diarrhea virus, which led to the loss of a significant number of piglets in 2013–2014 and an overall reduction in the amount of pork available that year. The outbreak led to even more stringent bio-security measures, including involving trucks in the effort. Farms had long used clothing changes and disinfectant to keep pigs safe, but now the truck drivers who transport the pigs, and even those who deliver food, are being required to take these same precautions. It adds about five hundred dollars in additional cost per truckload, but curtailing the spread of disease is worth it.[10]

The USDA has long banned the importation of meat, both fresh and cured, from countries where disease went uncontrolled. It is only recently that the ban on cured meats from several regions of Italy was lifted.[11] Of course, since Americans often visit countries where disease is not controlled, many farms no longer allow visitors. Something as simple as dirt on one's shoe from a country with infected pigs could wreak havoc.

The diseases that cause the greatest concern among those who raise pigs are foot-and-mouth disease, hog cholera, and African swine fever (ASF), and for good reasons. They can be devastating. For example, in the early 1980s, ASF swept into Haiti from the Dominican Republic, probably introduced from Spain. ASF is a hemorrhagic fever spread by ticks. There is no vaccine and no cure, and it usually kills close to 100 percent of infected animals. In Haiti, ASF killed two-thirds of the country's 1.2 million pigs, but because the remaining pigs would be carriers of the disease, they had to be destroyed as well. The United States then supplied specially bred, disease-free pigs to the eight hundred thousand peasant farmers who had lost their animals to the epidemic.[12]

People worldwide are protecting the genetic diversity of livestock to make sure we can keep a variety of breeds available, including disease-free or disease-resistant breeds. Organizations such as England's Rare Breed Survival Trust and the United States' Livestock Conservancy raise herds of heritage animals or collect semen to make certain that animals can be reproduced. But the real problem is one of numbers. The United States could replace all the pigs in Haiti, but who could replace all the pigs in the United States? It would have a catastrophic impact on the food supply worldwide. And potentially even more of an issue, how would we dispose of truly vast numbers of dead pigs?

Dr. Elisabeth Stoddard, whose areas of expertise are environmental studies and sustainable food systems, has studied the potential impact of an epidemic among swine in North Carolina, one of the biggest pork-producing states. While that epidemic has not happened as yet, there was a foretaste of the potential problems in 1999 when Hurricane Floyd hit North Carolina, drowning tens of thousands of pigs, as well as millions of chickens and turkeys.[13] North Carolina has, of course, had pigs since settlers first arrived in the 1600s. However, in the 1980s, as sales of the state's traditional top crop, tobacco, declined, farmers began turning to swine. Today, there are between nine and ten million hogs on North Carolina's flood plain. Population density and weather systems have created problems not faced in the Midwest but have given researchers like Stoddard a chance to consider what to do if there were a major epidemic.

Calling the industry "too big to fail," Stoddard explains that approximately 625,000 hogs are moved across the country every day, which, despite biosecurity measures at the farms, creates a risk of exposure to contagion. She explains that adult pigs can survive diseases like porcine epidemic diarrhea, but if a hog gets foot-and-mouth disease (FMD), international trade regulations make destroying the animal the most likely option. "Foot-and-mouth disease is the most contagious disease in veterinary medicine," Stoddard explains.

> It can even survive in bacon for a period of time, or leather, because cattle get FMD, too. So, tourists visiting from countries where the disease is common

> could easily introduce it. Animal scavengers might eat an animal that died from the disease and then walk across a border. An outbreak of FMD would mean stopping all movement of pork, both live and as products. The number of carcasses that we'd need to dispose of could be overwhelming, plus the disease could spread to other cloven-hoofed animals. North Carolina doesn't have enough places to bury the destroyed animals, largely because the ground water level here is so high. It could contaminate the water supply. We can't just burn the animals, as incineration can spread the disease. Composting is environmentally responsible, but hurricanes can make that less viable.

Stoddard emphasizes that, as for the farmers, there are no "bad guys" in this situation. There are simply people trying to earn a living while also trying to meet the growing market demand for affordable meat—and at the same time, trying to find a way to handle worst-case what ifs. Stoddard concludes, "Personally, I would like to see a shift toward farming practices that have been shown to reduce the risk of the spread of disease, as well as helping reduce other health and environmental issues that North Carolina faces because of the large number of pigs concentrated on the flood plain."

Bio-security may be a key element of our defense system, but faced with the possibility of disaster, farmers are taking prevention very seriously. "No entry" signs and strict limits on who has access to barns are the obvious indications of efforts to prevent the introduction of disease. Dr. Steve Henry relates that showers and clothes being changed before farmers, workers, and even doctors enter the barns is the norm today. However, it's not where the battle stops. "We're breeding pigs now that have disease resistance built in. Certain pathogens have actually been eliminated from pig populations. We can cut mortality by 30–40 percent." Henry relates that we are transitioning out of the age of antibiotics, and notes that farmers who have grown up using antibiotics are going to have to adjust. Henry states, "The big problem is disseminating the information, as almost everyone currently in agriculture has never worked in a system that didn't rely on antibiotics."

While some consider genetic modification problematic, many researchers see it as another weapon in the arsenal of those fighting disease. Research being done by Randall Prather at the University of Missouri and Bob Rowland of Kansas State University led to the discovery that removing a specific gene from pigs prevents their getting porcine reproductive and respiratory syndrome, a virus that accounted for about ten billion dollars in losses for the American pork industry in the last two decades. It is hoped that further research can find ways of preventing other diseases, but even stopping this one disease is a major triumph.[14]

Photo 14.1. Bio-security is an important consideration for big pig-raising operations. Signs like this one in South Dakota warn visitors to stay away from the barns, to keep pigs safe. Photo by Cynthia Clampitt.

Worth noting is that, while people are still working hard at keeping pigs healthy, they're also working on keeping humans healthy. Thanks to continued improvements on farms and in food processing, there has been a steady decline in food-borne illnesses associated with meat and poultry. The USDA and Centers for Disease Control keep close watch on all aspects of getting food from farm to consumer. So, while it is important to know and support efforts to keep livestock well, if we ourselves use reasonable safety precautions (keeping meat refrigerated, washing our hands after handling raw meat, cooking things to an appropriate temperature), we do not have to worry about the food that is being delivered to us.[15]

Hog Wild: The Growing Problem of Feral Pigs

When pigs get away from farms, they revert quickly to wild behavior. They actually become wild animals again. In addition to recent escapees, there are substantial populations of long-term wild pigs that are the descendants of animals dating to colonial days, plus a number of Eurasian wild boars that were imported for hunting reserves that have escaped or been released. Whether one calls them razorbacks, feral hogs, wild pigs, or wild boars, they are now a serious problem in forty states. And they are a costly problem, with an annual price tag of $1.5 billion in damage and control costs.[16]

Not surprisingly, the speed with which pigs reproduce is a liability in this situation. In fact, they reproduce faster than they can be hunted or trapped. And they're not all that easy to hunt, as feral pigs can run thirty miles per hour and jump a three-foot fence. Feral pigs destroy fields and forests, and kill and eat both indigenous wildlife and livestock, taking lambs, kids, calves, fawns, quails, newly hatched young of ground-nesting birds, frogs, crabs, mussels, snakes, and the eggs of endangered sea turtles. In challenges for territory, they have killed black bears. Wild pigs spread a wide range of diseases that can harm domestic livestock, wildlife, and even humans, including tuberculosis, bubonic plague, anthrax, FMD, and swine brucellosis. They can smell food at a distance of seven miles—and if they smell it, they'll find it. In Texas, where there is no winter to limit breeding, and where wild pigs date back to the Spanish conquest, numbers have topped 2.5 million, and the damage the wild pigs are doing is staggering.[17]

It's hard to know how many wild pigs there are, because they're very good at hiding. Estimates run as high as six million. When rooting, they can dig as much as three feet deep, which can completely destroy a nature reserve, recreational area, or farm. Even wild pigs still love corn, and they have been found to very methodically go down freshly planted rows, picking out every corn seed in a field. When wallowing, they often wreck water sources and streams, which further hurts both wildlife and the environment. They also root up or destroy native plants, leading not only to the loss of plants but also of habitats and food for native wildlife.[18]

States with wild pig problems allow and even encourage hunting. However, as prolific as the wild pigs are, if the population is reduced by 70 percent, it can rebound within two or three years.[19]

Even dead, wild pigs can be a problem, if hunters don't take necessary precautions. A significant number of wild pigs have brucellosis, which can be passed on to humans if the pigs are not cleaned properly before being consumed. But the real danger is having the disease get into the domestic pig population. Doing all that can be done to keep wild and domestic pigs apart is the only real protection against infection.[20]

Most states have now outlawed the importation of any additional Eurasian wild boars. Some states are even concerned about heritage breeds, as free-range animals can sometimes escape and are already closer to being wild than confinement pigs. A few years ago, Michigan considered outlawing heritage pigs, but instead mandated nose rings and more carefully tended fences. But the state also outlawed the introduction of any new wild boars, since many of the wild pigs in the state were introduced by those who thought they would be fun to hunt. Today, all but four of Michigan's eighty-three counties have reported feral pig sightings. As a result, the state has declared a year-round open season policy for hunting feral pigs, hoping they can prevent the kind of devastation some other states have experienced.[21]

Officials at the USDA have decided that a coordinated and collaborative effort is needed to keep the wild pigs from completely overrunning the country. In 2014, Congress gave the USDA Animal and Plant Health Inspection Service Wildlife Services twenty million dollars to begin a national feral swine management initiative. The hope is that they can both prevent the further spread of wild pigs and reduce the numbers of wild pigs now tearing up so much of the country.[22]

Fortunately, even though feral pigs are still a massive problem, some states are beginning to have success. Kansas, for example, has been really aggressive in its approach—something that is vital to a state that has Oklahoma, a state with a big feral hog population, as a neighbor. Biologists are determined to keep disease-carrying feral pigs from overrunning the state. Helicopters hovering over the wide, open plains give shooters an advantage. The success Kansas has had thus far in reducing numbers has been applauded by biologists in other states.[23] One can hope that this success is reproduced elsewhere.

The People Problem

Worldwide, there are too many to feed and, in increasingly urbanized countries, too few to provide the food. More and more people are leaving behind rural settings and heading for less strenuous, higher-paying jobs in the cities. This means that at the time we most need people raising food, there are fewer people interested in doing it.

Urban populations in developed nations, such as the United States, tend to be further removed from a rural past than those in developing countries. Most city dwellers have little concept of what is involved in raising livestock and may have visions of more muck than they want to consider. However, for those who might be interested, know that not all farm jobs will get one's boots dirty.

"In the industry, we're seeing a demand for more technological training," Dr. Steve Henry relates.

> So much is now monitored by computer—the feed, the water, the temperature, but also how the pigs are behaving. The systems are highly integrated and track all aspects of the operation. So, we need more people who can work with the technology. However, we also need hands-on people, people who help the sows farrow, who rescue the runts that are being attacked, that check the health of the pigs daily, and even simply keep the pigs accustomed to humans.

Henry continues,

> The rest of the world understands what is at stake. We're exporting these technologies to China and South America. They want what we know. I've been to South America a few times now, and I've seen how eager they are to change, to adopt successful methods of feeding their growing populations. Chile, for example, has not only adopted the latest farming technology from the U.S., they've also brought in hydrology experts from Israel and breeding experts from Spain. They've taken the best of everything and do it better than we do, because in countries where a lot of people are hungry, there is more motivation to participate in solving the problem.

The problem that needs to be solved is hunger. The world population in 1950 was 2.5 billion people. By 2005, it had risen to 6.5 billion, and it reached 7.5 billion in early 2018. That's a stunning growth rate. Most of the growth has been in less-developed countries, which generally means countries that have not developed extensive systems for producing and delivering food.[24]

With five billion more people who need to be fed appearing within a very short period of time, it's not surprising that the world ended up with a hunger problem. However, we have actually managed to make remarkable headway in reducing the number of people who are hungry. In 1990, the United Nations set a twenty-five-year goal of reducing by half the number of people who did not have enough food. Despite continued population growth, the goal was achieved. The percentage of the population in developing nations that gets insufficient food has dropped from 23.3 percent to 12.9 percent.[25]

In addition to trying to find more workers, another way of dealing with the need to produce more food is through breeding programs. Henry relates, "Genetically, we've made great headway. Sows are now producing more healthy piglets, which translates to more protein going to market. In 1976, the number of surviving piglets a sow had meant we sent less than 2,000 pounds live weight of meat to market. Today, a sow can send more than 6,000 pounds live weight of meat to market from her offspring."

The planet cannot support unlimited population growth, but even if growth levels off, we'll still always need people to raise the food.

Care Conundrums

While some of the issues related to pigs are unarguably problems, some are a matter of differing opinions. A considerable amount of controversy these days revolves around how pigs are raised. As discussions with different farmers in this book have shown, every option has benefits and drawbacks. USDA guidelines show that pork is safer now because pigs are being raised indoors, but large confinement operations are less natural for pigs (though perhaps less unnatural than city streets or dumps). A more idyllic life and, in the case of heritage breeds, more flavorful meat are afforded by open range operations, but these pose danger to the pigs in the form of predators and create worry about erosion, pollution, and escaped pigs going feral, plus they can't provide enough pork to meet even current demand.

We've also seen that the treatment of pigs in the United States has improved steadily, advancing from viewing them as mobile garbage disposal units to genuine concern for their well-being. Intentional abuse is rare, as most farmers value their animals, but even those who don't would find it difficult to sell meat from animals that are bruised or injured. In recent years, there has been increased focus on reducing stress and keeping pigs happier.

Some of the concern expressed in popular culture is due to the lack of knowledge of what is actually involved in farming. Sarah Schieck, Swine Extension Educator with the University of Minnesota Extension, notes as an example that people frequently protest hormone use, but there are no hormones used in raising pigs. There never have been. Schieck explains that no hormones have even been approved for use with pigs. There's no need. Pigs are designed to gain weight.

Lately, pork processors and food companies have begun requiring more information from the farmers who provide the pigs they process. These companies use audits to show that they are being proactive. Schieck relates, "Most of what is in the audits that these companies demand was already being done. It's just that most people are completely separated from farming, so they tend to believe inflammatory ads that suggest bad practices. A packer will call and announce, 'We're coming out to audit your pigs,' and the farmer has to stop working to accommodate the audit." One of Schieck's roles is conducting Common Swine Industry Audit workshops that help farmers prepare for an audit. "Previously, several of the major packing companies had quite different audits," Schieck states. "For a farmer who sells to more than one packer, that became hugely time consuming, keeping farmers in the office rather than with their pigs. Now, most big processors have accepted the common audit, which simplifies the process. This helps farmers in another way, because farmers are required to keep daily records of how much time they spend with their pigs. Audits take away from that time."

Schieck also notes that decisions, both past and present, about raising pigs depend on many factors, from the strengths of farmers to available resources to how pigs behave. Farrowing and breeding are the most labor-intensive stages, and younger pigs are more susceptible to disease, so more workers and more care are needed. As pigs approach the finishing stage, more feed is needed. (This is why, even when piglets are born in North Carolina, where warmer weather is good for the newborns, older pigs often get shipped to the Midwest for finishing, as there is vastly more food available.) However, less moving reduces stress. There's an adjustment period every time pigs are moved, which is why wean-to-finish operations are becoming more common. When pigs do have to be moved, Schieck relates, an "all in/all out" management approach is often employed, where an entire generation is moved together. This reduces the introduction of disease by mixing pigs.

Worth remembering is that pigs fight. "Much depends on the farmer and the number of workers," Schieck continues. "Pigs will choose one animal to pick on, and then they'll keep attacking it. Animals often need to be rescued, to minimize damage or avoid death." Fighting is just one of the reasons workers need to spend a lot of time with pigs, and it's one of the reasons some types of enclosures are still needed, at least for some pigs.

That said, many pig-raising operations have moved away from gestation crates, or are in the process of moving away. The crates are used to keep pregnant sows from fighting. Several large corporations and even some states have taken a stand against the crates, also known as sow stalls, where sows are confined for the 114 days of their pregnancies. Still, the issue is not settled, as there are reasons to use the crates. In a world where getting good farm workers is often difficult, it takes fewer workers and less skill to manage sows if they are in crates. Science has established that there are no negative psychological effects and few physical problems. However, the idea of having an animal in a crate that allows little movement seems uncomfortable and unkind. Farrowing crates are less controversial, since they save the lives of so many piglets and sows are in them for a shorter time. Still, it is not a settled issue everywhere.

Megan Kelly is a swine technician and has worked on both average and very large hog farms. During college, she also had the opportunity to be a meat judge, after studying meat evaluation. She has a bachelor of science degree in agriculture from Ohio State, majoring in animal science, with a focus on animal industries—a level of education that is increasingly common among American farmers. She is enthusiastic about livestock raising and concerned that so many misconceptions exist about the industry.

Kelly says that when people she meets learn about her work, they bring up all the rumors they've heard.

> People worry that animals are not treated humanely. During my experience on the Meats Judging Team, I learned that when carcasses are graded, the carcasses that have bruises and damage to the muscle actually have their value decreased. So even if owners didn't care about the animals, though most do, it is definitely in their best interest to treat the animals well. Otherwise, they're losing money. Employers do a lot of educating of employees and enforcing of proper animal handling, because they know that animals that are healthy and happy bring in the most money.

The perception, often presented by advertising, that food is produced by factories is another thing that Kelly finds distressing. There are big farms and even corporate farms, but most farms in the United States, even most of the really large ones, are family farms. Yes, they are still businesses, but they are family businesses.[26]

Addressing the issue of farrowing crates, Kelly notes,

> These crates ensure the safety of the sow, her piglets, and the employees who work in the barn. As one of the people in there with the sow, a swine technician appreciates this aspect in particular. As with any species of animal, mothers are often protective, and these sows weigh from 400–600 pounds. There have been many times when I would have gotten severely injured had that sow not been in a farrowing crate.

Kelly explains that the crates, which are actually just a few bars that limit the sows' movements, still allow the piglets full access to mom. In fact, the piglets can move around as much as they want. They can easily feed, which is their main concern as newborns. But they can also escape mom, when she gets tired of being bothered. Kelly continues,

> Sows have a tendency to crush their piglets by laying on them. During farrowing, we're in there, circulating among pens, making sure piglets are safe. These crates have substantially lowered the number of piglet deaths by crushing. The crates allow the sow to stand up, feed, and lay down on either of her sides, so she can get comfortable. In the open space for piglets, there are either heat lamps or heated mats to keep the piglets warm and dry during their first days, because piglets want to be warm, but sows want to be cool after farrowing. In addition, young mothers sometimes get aggressive with their piglets. This behavior is known as "savaging." The crates help keep the piglets away from the mother's potentially fatal attacks.

Some farms also have gestation crates, Kelly points out. These are larger than farrowing crates, so the sows can move around more, but they keep the sows apart. Kelly relates,

> Sows have a hierarchy, and they can cause damage to themselves and others establishing that hierarchy. I've worked in a barn with an open pen and one that used gestation crates. I witnessed the pros and cons of each. In the open pen, sows were more used to being handled, as people have to get into the pens to care for individual pigs. However, there were more problems with smaller or weaker sows being blocked from getting to food or getting beat up by more aggressive sows, resulting in injuries, due to dominance establishment. With crate gestation, the sows are safe, but they were harder to handle and were more aggressive toward us. So, there is no easy, obvious answer.

Extension Educator Sarah Schieck points out that most large meat packers now consult handling experts, such as Temple Grandin, to make certain stress is minimized as pigs pass through their operations. "Facilities are designed to reduce the transmission of disease," Schieck notes,

> but they're also designed to keep pigs comfortable. To reduce heat stress in summer, there are fans, cool mats for sows, dripping water that pigs can access—and the focus is on doing it without using many fossil fuels. Also, food is especially formulated to meet the needs of pigs. It's clean and safe. Farmers work with swine nutritionists to ensure that the animals get all the vitamins, minerals, protein, and carbs that they need. Farmers want them to be strong, so they want them to eat well. Most pigs eat better than most humans.

Even transportation is vigilantly monitored. Schieck, who also works as a transport quality assurance advisor, relates that truck trailers are carefully prepared to care for pigs. They are kept clean, warm in winter and cool in summer, and the drivers must be Transport Quality Assurance certified to work, with recertification required every three years. This is a national program, so all states require this level of care.

Schieck concludes emphatically, "Farmers have always been concerned about their animals. People are continually doing research and updating practices, and university extension programs make certain farmers have the latest information, so they can continue to do a good job. We just need to communicate that better."

Experimental Pig

On the flip side of the pig health coin, there is one very important advantage to their being so similar to humans in regards to the maladies from which they can suffer, and that is that we can do things with pigs that can help humans. But can we go too far?

Today, a great deal of research is being done with pigs. Pigs have several internal systems that work very much like those in humans, and heart valves, for example, have been successfully transplanted from pigs into humans. Pigs have been used for more than three decades to help search for ways to solve human health problems. Before synthetic insulin was created, diabetics received pig insulin. Because they are omnivores, like we are, pigs process nutrients in a similar manner, so they're useful for dietary studies. One dramatic recent discovery was regrowing human leg muscles using implants of pig bladder tissue.[27]

Because of similarities to human systems, we can use pigs to discover things about ourselves. Almost everyone today knows about endorphins, but they are a relatively recent discovery. In 1975, John Hughes and Hans Kosterlitz published their discovery of a small, morphine-like amino-acid molecule that they had found in the brains of pigs. They had actually spent a fair bit of time looking for what they initially called Substance X, but after examining a lot of pig brains, they found their molecule. It was another scientist, Choh Hao Li at the University of California in San Francisco, who tested it and found that it did act like morphine—it killed pain and was addictive.[28] (*Endorphin* was coined from the words *endo*, meaning "inside or within," combined with *morphine*.)

Pigs are being used for everything from drug tests to transplantable parts for humans.[29] Despite the obvious advantages, however, there is concern that we can go too far. Pigs have been cloned, in order to improve disease resistance. Some are discussing the possibility of genetically engineering pigs to make it easier to use them for replacement parts for humans. The debate will probably continue for some time as to where lines should be drawn, but one can hope that, whatever the outcome, it does not have negative consequences.

International Entanglements

Globalization is here to stay. It is something that offers both challenges, including the possible spread of disease, and hope, from improved opportunities overseas to feeding growing populations everywhere. People, products, and germs are never more than a few hours away with air travel, or a few days by sea. Tourism can pose a threat in the form of everything from dirt innocently brought in on shoes to people intentionally violating prohibitions on bringing certain products (meats, plants, animal skins, etc.) into the country. But tourism is still a drop in the bucket compared to the massive, international import/export trade.

European imports have become a sign of affluence around the world. *Prosciutto di Parma* is one of Italy's top exports, appearing in restaurants and gourmet shops worldwide, and the popularity of European salami and hams is growing in increasingly affluent nations. But a far greater amount of meat being moved is not cured. Chilled or frozen pork is shipped by the millions of pounds, and the amount is increasing rapidly.

Japan is the largest importer of pork products in the world, with half of all the pork the country consumes coming from Europe (Denmark and Germany being the number one and number two European sources) and North America, but an increasing amount is coming from China, generally in the form of sausage. Other top pork importers include Mexico, China, Russia, and South Korea. Interestingly, though Canada has increased its production of pork in recent years, it tends to sell that pork to foreign markets, where prices are higher, and then imports pork from the United States for local consumption.[30]

While processing, chilling, and freezing of meat make more food available for more people, a fear of spreading disease keeps the international market for pork from expanding any faster than it has.[31] The 2013 outbreak of porcine diarrhea epidemic, with its similarity to the Chinese version of the virus, has people looking ever closer at how to keep disease from jumping continents. The USDA ban on imports from some regions of Italy was mentioned above, but Italy is not the only country with banned meat—and disease is not the only reason for bans. Russia has on occasion instituted bans for political reasons, such as limiting U.S. imports when we disagreed over the treatment of Ukraine.[32] The impact of political bans is perhaps not as obvious as that of stopping disease from spreading, but it can lead to people going hungry—or looking to less-safe sources of food.

Until recently, in China, home to roughly half of the planet's one billion domestic pigs, many of the country's 1.4 billion people simply had a few animals at home. With a few pigs per family, they really didn't eat all that much meat. However, in recent years, there has been a flood of people moving from China's rural areas into the country's cities, looking for higher-paying jobs. The double dilemma is that better jobs and higher incomes have led to greatly increased pork consumption, while at the same time, far fewer people are raising pigs. Demand has increased on average 5.7 percent per year. Since pork consumption in China is more than six times greater than U.S. consumption, that increase represents a substantial amount of pork.[33]

Because large-scale centralized processing was not traditionally a consideration in China, the country has struggled to gear up production to meet the greater demand, with often seriously problematic results, from criminal gangs selling diseased pigs to dead pigs being dumped by the thousands in local rivers.[34]

The Chinese government has begun to institute laws to reduce overcrowding on farms, limiting the number of pigs farmers can raise. As a result, importation of pork from the United States has spiked upward. But China still needed more meat. In September 2013, Shuanghui International Holdings Limited, a majority shareholder in China's largest meat processor, bought the world's largest pork producer, Virginia-based Smithfield Foods.[35] In addition to China's need for pork, the sale was also prompted by a general mistrust in China of locally produced pork.

Smithfield's stockholders did well, and the company's sales have increased, fueled at least in part by the success of China's efforts to limit the number of their pigs. In the first six months of 2015, pork exports to China grew by 45 percent.[36] There are those who are worried about the situation, as China has had so many problems, but others hope that China having access to Smithfield's considerably greater technology and research will have a positive effect on China's meat production. Another question being asked, as demand in China increases, is how many more pigs can the United States raise before they have an impact on our environment. However, concern is not one-sided. The Chinese are hoping to eventually build their own advanced pig farms. The debates continue. It is one more thing that needs to be monitored closely, as the numbers are so great that the impact could be immense.[37]

China is not alone in the dilemma of feeding its people. As developing countries build stronger economies and people have the opportunity to earn more money, they all face the issue of having people moving to cities, having higher incomes, and wanting to eat at better than subsistence level, which means more protein. These are also the countries that are seeing the most population growth, so demand will continue to increase. The pressure to devise ways to raise more food with fewer farmers, generally on less land, is something we are all going to have to deal with—not just raising more to send overseas, but developing technologies and methodologies that we can export, as well, so that countries can feed themselves.[38]

Reality and Hope

So, there are problems and challenges, but there are also people working on resolving them. There is never going to be a time when people can stop caring or trying or working, but it is encouraging to see that the care and work are having results, from protecting the environment to preventing disease.

Not everyone agrees on what is being done, but that is true of any endeavor. Differences of opinion are to be encouraged, as they create dialogue and help propel forward movement. However, the opinions need to be grounded in what is really happening, not just here, but everywhere. Globalization means we are operating in a larger context.

And don't believe anyone who offers an easy solution. There isn't one.

Fortunately, in the United States, there are people at every level of the system who care deeply about everything from world hunger to the welfare of livestock. It is important to remember that, while different American farmers may have different goals, as well as different pigs and farms, they all have in common the desire to keep pigs healthy and consumers happy, and most are also eager to reduce hunger wherever it occurs.

Everyone is trying to do what is right. Farming in particular is too much work and not enough money to do it for any reason other than believing one is doing something good.

Notes

Introduction

1. United Nations, Food and Agriculture Organization, http://www.fao.org/ag/againfo/themes/en/meat/background.html.

2. Brett Mizelle, *Pig*, p. 7.

3. James R. Shortridge, *The Middle West*, pp. 20–21.

4. Jon K. Lauck, *The Lost Region*, pp. 21–23.

5. Pork and beef were not the only sources of protein. Game on the frontier and fish near waterways were also common. However, traditions brought from Europe made meat the dominant protein source, and pork and then beef the dominant meats. Chicken was not a major food source for Europeans through most of history (see City Chicken recipe in chapter 10 for more details). However, in this book, the discussion will focus on meat, which the U.S. Department of Agriculture (as well as most cookbooks and menus) defines as mammals, with chicken being poultry. So, when I speak of meat, I'm using the U.S. Department of Agriculture designation.

6. Katharine M. Rogers, *Pork: A Global History*, p. 84.

7. Stephen J. Dubner, "Beef or Chicken? A Look at U.S. Meat Trends in the Last Century," *Freakonomics*, December 9, 2010.

8. North American Meat Institute, "The United States Meat Industry at a Glance," http://www.meatami.com/ht/d/sp/i/47465/pid/47465.

9. Megan Durisin, "It's Pork over Beef in America for First Time Since 1952," *Bloomberg Business*, February 2, 2015. Lydia Mulvany, "Beef Isn't For Dinner Anymore as Americans Devour Cheaper Pork," *Bloomberg Business*, October 4, 2015.

10. From a speech by A. Richard Crabb, author of *The Hybrid-Corn Makers*, given on October 7, 1944, at a meeting of the State Historical Society, from the archives of the McLean County Historical Museum.

11. Daniel W. Gade, "Hogs (Pigs)," *The Cambridge World History of Food*, p. 537.

12. Christopher L. Delgado, "Rising Consumption of Meat and Milk in Developing Countries Has Created a New Food Revolution," *The Journal of Nutrition*, vol. 133 no. 11, November 1, 2003.

Chapter 1. Meet the Pig

1. Caroline Grigson, "Culture, Ecology, and Pigs from the 5th to the 3rd Millennium B.C. Around the Fertile Crescent," in *Pigs and Humans: 10,000 Years of Interaction*, p. 98.

2. Michael Fisher, "Prevent Swine from Getting Sunburn," *High Plains/Midwest Ag Journal*, June 11, 2011.

3. Celia Lewis, *The Illustrated Guide to Pigs*, p. 14

4. Lyall Watson, *The Whole Hog: Exploring the Extraordinary Potential of Pigs*, pp. 25, 26.

5. These figures are approximate, with the actual consumption affected by what is being fed and the type of pigs being kept. Malcolm DeKryger, who supplied these figures, raises pigs bred to meet U.S. consumer demand for leaner meat. Celia Lewis, in the *Illustrated Guide to Pigs*, pp. 14, 24, offers slightly lower figures, but she is describing the feeding of hand-raised heritage breeds on small farms.

6. Watson, p. 16.

7. "Dental Anatomy of Pigs," Colorado State, http://www.vivo.colostate.edu/hbooks. Watson, p. 31.

8. Marco Masseti, "The Economic Role of *Sus* in Early Human Fishing Communities," in *Pigs and Humans*, pp. 161, 162, 164–66. Watson, p. 98.

9. Temple Grandin and Catherine Johnson, *Animals Make Us Human: Creating the Best Life for Animals*, p. 176.

10. Masseti, p. 169.

11. Lewis, p. 25. Jared Diamond, in *Guns, Germs, and Steel: The Fates of Human Societies*, suggests that influenza may have arisen in China, since pigs, which transferred the disease to humans, were so important so early in Chinese history; p. 330.

12. Masseti, pp. 161, 169. Daniel W. Gade, "Hogs (Pigs)," in *The Cambridge World History of Food*, p. 537.

13. Watson, p. 47.

14. Damage done to farms and use of rings in pigs' noses on the American frontier: Nicholas P. Hardeman, *Shucks, Shocks, and Hominy Blocks*, p. 197. Laws during Middle Ages to use "rings," actually curved bits of wire: Frances and Joseph Gies, in *Daily Life in Medieval Times*, p. 192. Not interfering with foraging: "The effect of nose rings on the exploratory behaviour of outdoor gilts exposed to different tests," by Merete Studnitz, Karin Hjelholt Jensen, and Erik Jørgensen, *Applied Animal Behavior Science*, vol. 84 no. 1, November 3, 2003, pp. 41–57.

15. Grandin and Johnson, pp. 173, 184, 185.

16. Grandin and Johnson, p. 179. Temple Grandin and Catherine Johnson, *Animals in Translation*, pp. 154–56, 170.

17. Keith Dobney, Anton Ervynck, Umberto Albarella, and Peter Rowley-Conwy, "The chronology and frequency of a stress marker (linear enamel hypoplasia) in recent and archaeological populations of *Sus scrofa* in north-west Europe, and the effects of early domestication," *Journal of Zoology*, vol. 264 no. 2, February 2004, pp. 197–208. Margret Tuchsherer, Birger Puppe, Armin Tuchscherer, and Ellen Kanitz, "Effects of social status after mixing on immune, metabolic, and endocrine responses in pigs," *Physiology and Behavior*, vol. 64 no. 3, June 1, 1998, pp. 353–60. Birger Puppe, "Effects of familiarity and relatedness on agonistic pair relationships in newly mixed domestic pigs," *Applied Animal Behaviour Science*, vol. 58 no. 3–4, July 1998, pp. 233–39. J. Elizabeth Bolhuis, Willem G. P. Schouten, Johan W. Schrama, and Victor M. Wiegant, "Individual coping characteristics, aggressiveness and fighting strategies in pigs," *Animal Behaviour*, vol. 69 no. 5, May 2005, pp. 1085–91. Grandin and Johnson, *Animals Make Us Human*, pp. 182, 184, 185; Grandin and Johnson, *Animals in Translation*, p. 104.

18. Lewis, p. 27.

19. Watson, p. 98.

20. "Dental Anatomy of Pigs," http://www.vivo.colostate.edu/hbooks/pathphys/digestion/pregastric/pigpage.html.

21. Lewis, p. 10.

22. Lewis, pp. 10, 11.

23. Watson, p. 106.

24. Brett Mizelle, *Pig*, p. 7.

25. Gade, "Hogs (Pigs)," p. 537. Watson, p. 105.

26. Watson, p. 102.

27. Grigson, "Culture, Ecology, and Pigs," p. 98. Grigson states that pigs "are a natural species, and there is no reason to think that their basic biology has been changed by domestication *per se*."

28. Watson, p. 102.

29. Watson, p. 25.

30. Lewis, p. 19.

31. There are numerous lists available. Here's one from CBS News: http://www.cbsnews.com/pictures/natures-5-smartest-animal-species/.

32. Natalie Angier, "Pigs Prove to Be Smart, If Not Vain," *New York Times*, November 9, 2009.

Chapter 2. Early Pig: Domestication and Early Civilization

1. Jared Diamond, *Guns, Germs, and Steel: The Fates of Human Societies*, p. 86.

2. Umberto Albarella et al., *Pigs and Humans: 10,000 Years of Interaction*, pp. iii, 1, 3, 5.

3. Lyall Watson, *The Whole Hog: Exploring the Extraordinary Potential of Pigs*, pp. 91–92.

4. Albarella et al., p. 3.

5. Hallan Cemi Tepe, *The Megalithic Portal*, http://www.megalithic.co.uk/article.php?sid=15001.

6. Brian L. Peasnall, "The Intricacies of Hallan Cemi," *Expedition Magazine*, vol. 44 no. 1, March 2002.

7. Keith Dobney, Anton Ervynck, Umberto Albarella, and Peter Rowley-Conwy, "The Transition from Wild Boar to Domestic Pig in Eurasia, Illustrated by Tooth Developmental Defect and Biometrical Data," in Albarella et al., *Pigs and Humans*, pp. 57, 66, 69. Watson, pp. 94–96.

8. Watson, p. 95.

9. Peter Kaminsky, *Pig Perfect: Encounters with Remarkable Swine and Some Great Ways to Cook Them*, p. 59.

10. Jared Diamond points out in *Guns, Germs, and Steel* that a few cultures had such abundant resources that they were able to settle without agriculture, at least initially, the primary examples being Japan and the Pacific Northwest, where the sea supplies abundantly. As a result, people could survive for a while without raising crops or domesticating animals. However, as populations grew, agriculture and domestic animals became necessary.

11. Watson, pp. 92, 94.

12. Diamond, *Guns, Germs, and Steel*, p. 167.

13. Greger Larson, Ranran Liu, Xingbo Zhao, Jing Yuan, Dorian Fuller, Loukas Barton, Keith Dobney, Qipeng Fan, Zhiliang Gu, Xia-Hui Liu, Peng Lv, Leif Andersson, and Ning Li, "Patterns of East Asian pig domestication, migration, and turnover revealed by modern and ancient DNA," *Proceedings of the National Academy of Sciences of the United States of America*, vol. 107 no. 17, October 28, 2009.

14. Daniel W. Gade, "Hogs (Pigs)," in *The Cambridge World History of Food*, p. 537.

15. Caroline Grigson, "Culture, Ecology, and Pigs from the 5th to the 3rd Millennium B.C. Around the Fertile Crescent," in Albarella et al., *Pigs and Humans*, p. 83.

16. "Domestic Animals in Harappan Levels," Shodhganga, Chapter 3: Archaeological Evidences, p. 56. http://shodhganga.inflibnet.ac.in/bitstream/10603/7817/10/10_chapter%203.pdf.

17. The Ancient Indus Civilization, http://www.harappa.com/har/indus-saraswati.html.

18. "Domestic Animals in Harappan Levels," pp. 45, 91, 112, 113.

19. Rachel Laudan, *Cuisine and Empire: Cooking in World History*, pp. 36–37.

20. Karen Rhea Nemet-Nejat, *Daily Life in Ancient Mesopotamia*, p. 159.

21. Watson, p. 129. Watson also explains, on p. 46, that pigs are good at finding truffles because black truffles produce a chemical that copies the testosterone normally found in a boar's salivary glands, which pigs can smell.

22. Barbara Mertz, *Red Land, Black Land: Daily Life in Ancient Egypt*, p. 108.

23. Desmond Whyman, *Shoulder of Mutton Field*, p. 39, Nottingham University Press, 2010, discusses the difficulty of identifying who might have made sausage first, as well as the attribution to Assyrians in Tyre. Whyman also notes that one of the first written descriptions of blood sausage was in Homer's *Odyssey*, which dates to 850 B.C., so clearly an ancient creation. Prosper Montagné, *The New Larousse Gastronomique*, p. 719, is more absolute in his statement that blood pudding was an Assyrian invention, a specialty of the butchers of Tyre, and precursor of *boudin noir*. Katharine M. Rogers, *Pork: A Global History*, p. 25, also identifies pudding/sausage made of pig's blood as an Assyrian invention.

24. University of York, "Feasts and Food Choices: Culinary Habits of the Stonehenge Builders," *Science Daily*, October 13, 2015.

25. Leviticus 11.

26. Sarah Phillips, "The Pig in Medieval Icongraphy," in Albarella et al., *Pigs and Humans*, p. 383.

27. http://www.adfg.alaska.gov/index.cfm?adfg=disease.muscle2 and http://latimesblogs.latimes.com/outposts/2010/07/fish-and-game-q-and-a-1.html.

Chapter 3. Old World Pig: Pork and the Rise of Familiar Cultures

1. Peter Berresford Ellis, *The Celts: A History*, pp. 97–98, Mark Kurlansky, *Salt: A World History*, pp. 56–59. Jacqui Wood, *Prehistoric Cooking*, p. 47. Jacqui Wood, *Tasting the Past*, p. 9.

2. Strabo, *Geography*, Book IV, chapter 4, paragraph 3.

3. Smoking is likely the oldest form of preserving meat, probably starting close to the first cooking with fire. Smoke dries meat, though chemicals present in wood smoke also aid in preservation. However, smoke is not as effective at preserving meat as salt is. The main advantage to smoke was that everyone had access to it, whereas salt had to be obtained by hard work or trade. A surprising amount of the world's history has revolved around getting salt. For details on that pursuit, check out Mark Kurlansky's *Salt: A World History*. Today, in countries with adequate refrigeration, the continued use of salt and smoke are primarily for flavor.

4. Lyall Watson, *The Whole Hog: Exploring the Extraordinary Potential of Pigs*, p. 107.

5. James Mallory and Douglas Q. Adams, *Encyclopedia of Indo-European Culture*, pp. 426–27. Watson, p. 107. Ellis, pp. 102–03.

6. I. G. Kidd, *Posidonius: Volume III The Translation of the Fragments*, Cambridge, UK: Cambridge University Press, p. 134.

7. Wood, *Prehistoric Cooking*, p. 90. Wood, *Tasting the Past*, p. 14. Wood includes recipes for many of the dishes suggested by archaeological evidence, and some of them sound pretty good.

8. Reay Tannahill, *Food in History*, pp. 65–69.

9. Veronika Grimm, "The Good Things that Lay at Hand," from Paul Freedman, *Food: The History of Taste*, pp. 70–71.

10. Katharine M. Rogers, *Pork: A Global History*, p. 25

11. Rachel Laudan, *Cuisine and Empire: Cooking in World History*, pp. 71–72.

12. Jacques Heurgon, *Daily Life of the Etruscans*, pp. 117–18, 185–90. Patrick Faas, *Around the Roman Table: Food and Feasting in Ancient Rome*, pp. 15–16.

13. Laudan, p. 71. Heurgon, p. 117. Faas, p. 3.

14. Evan D. G. Fraser and Andrew Rimas, *Empires of Food: Feast, Famine, and the Rise and Fall of Civilizations*, pp. 43, 47, 52–57. Neville Morley, "The Profits of Empire," in the *Cambridge Illustrated History of the Roman World*, p. 295.

15. Martin Jones, *Feast: Why Humans Share Food*, pp. 210–15. Fraser and Rimas, pp. 44–45. Laudan, p. 80.

16. Tannahill, p. 71.

17. Waverley Root, *Food: An Authoritative and Visual History and Dictionary of the Foods of the World*, pp. 371–74. Also described in Pliny the Elder's *Natural History*, Book VIII, "The Hog," which is chapter 77, p. 344 of the 1855 John Bostock translation.

18. Jones, p. 225. Wood, *Tasting the Past*, pp. 20, 24. Joseph Dommers Vehling, *Apicius: Cookery and Dining in Imperial Rome*, pp. 14, 15, 16, 30.

19. Vehling, pp. 16, 32, 70.

20. Faas, pp. 259–61.

21. Jones, pp. 212–15. Tannahill, pp. 78–80, 90.

22. Fraser and Rimas, pp. 16–17.

23. Fraser and Rimas, p. 61. Clyde Pharr, *The Theodosian Code and Novels, and the Sirmondian Constitutions*, Union, NJ: The Lawbook Exchange, Ltd., 2001, pp. xx.

24. Frances and Joseph Gies, *Daily Life in Medieval Times*, pp. 78–79, 123–24, 165–66, 192. David B. Danbom, *Born in the Country*, p. 6. Werner Rösener, *Peasants in the Middle Ages*, p. 97. William Chester Jordan, *Europe in the High Middle Ages*, p. 292.

25. Régine Pernoud, *Those Terrible Middle Ages: Debunking the Myths*, San Francisco, CA: Ignatius Press, 2000, pp. 7, 8, 12–14.

26. Gies, pp. 67–71. Nicola Fletcher, *Charlemagne's Tablecloth: A Piquant History of Feasting*, p. 120. Sarah Phillips, "The Pig in Medieval Iconography," Albarella et al., *Pigs and Humans: 10,000 Years of Interaction*, p. 375.

27. Watson, p. 118. Rösener, p. 135.

28. Jordan, pp. 182–83.

29. Watson, pp. 118–19. Rösener, p. 135. Anton Ervynck, An Lentacker, Gundula Müldner, Mike Richards, and Keith Dobney, "An Investigation into the Transition from Forest Dwelling Pigs to Farm Animals in Medieval Flanders, Belgium," Albarella etal., *Pigs and Humans*, pp 171–73.

30. Root, p. 372. C. M. Woolgar, "Feasting and Fasting: Food and Taste in Europe in the Middle Ages," from Freedman, *Food: The History of Taste*, p. 168.

31. John Scofield and Alan Vince, *Medieval Towns: The Archaeology of British Towns in Their European Settings*, Leicester University Press, 2003, p. 82–83.

32. Watson, p. 119.

33. Dolly Jørgensen, "Running Amuck? Urban Swine Management in Late Medieval England," *Agricultural History*, vol. 87 no. 4, Fall 2013, pp. 429–51.

34. Fraser and Rimas, pp. 20–29.

35. Laudan, pp. 178–82. Madeleine Pelner Cosman, *Fabulous Feasts: Medieval Cookery and Ceremony*, pp. 20–25, 198–203. Woolgar, pp. 168–70.

36. Mark Essig, *Lesser Beasts*, pp. 107–08. NOAA National Climate Data Center, http://www.ncdc.noaa.gov/paleo/globalwarming/medieval.html. Fraser and Rimas, pp. 30–35. Jordan, p. 293.

37. Jared Diamond, *Guns, Germs, and Steel: The Fates of Human Societies*, pp. 323, 332.

38. Daniel W. Gade, "Hogs (Pigs)," *The Cambridge World History of Food*, p. 538. Essig, p. 50. Rogers, pp. 91–92. Tannahill, pp. 40–41. Fletcher, pp. 122–23.

39. Diamond, pp. 331–33.

40. Laudan, pp. 2–3. Diamond, pp. 334–43.

41. Laudan, pp. 72–73. Tannahill, p. 105.

42. R. Behl, N. Sheoran, J. Behl, and R. K. Vijh, "Genetic Analysis of Ankamali Pigs of India Using Microsatellite Markers and their Comparison with Other Domesticated Indian Pigs," *Journal of Animal Breeding and Genetics*, vol. 123 no. 2, April 2006, pages 131–35.

43. USDA Foreign Agriculture Service Global Agricultural Information Network Report, India: Pork—Annual 2011, GAIN Report Number IN1128, 3/28/2011. Jagdish Kumar, "Can Pork Consumption Fix India's Cheap Protein Link," *The Pig Site*, May 18, 2015.

44. Rogers, pp. 20–21. Essig, pp. 60–63.

45. Jeffrey Yoskowitz, "On Israel's Only Jewish-run Pig Farm, It's the Swine That Bring Home the Bacon," *Haaretz*, Saturday, July 18, 2015, Av 2, 5775. Heather Sharp, "Israeli Pig-Farming Kibbutz Draws Religious Ire," *BBC News*, Wednesday, June 30, 2010.

46. Essig, pp. 102–03.

47. http://muslimheritage.com/article/alfraganus-and-elements-astronomy. Silvio A. Bedini, ed., *The Christopher Columbus Encyclopedia*, Basingstoke: Simon and Schuster, 1992, p. 51.

48. "The Round Earth and Christopher Columbus," https://pwg.gsfc.nasa.gov/stargaze/Scolumb.htm.

49. Julian Wiseman, *The Pig: A British History*, p. 2. Brett Mizelle, *Pig*, pp. 20–23. Allan G. Bogue, *From Prairie to Corn Belt*, p. 106.

50. Watson, p. 103. Rogers, p. 92. Wiseman, pp. 38–56.

Chapter 4. Colonial Pig: Pigs in the Age of Exploration

1. Reay Tannahill, *Food in History*, pp. 224–26.
2. "Treaty Between Spain and Portugal concluded at Tordesillas; June 7, 1494," translation, Lillian Goldman Library, Yale Law School, http://avalon.law.yale.edu/15th_century/mod001.asp.
3. Alfred W. Crosby, Jr., *The Columbian Exchange: Biological and Cultural Consequences of 1492*, pp. 75–76.
4. Mark Essig, *Lesser Beasts*, p. 120. Tannahill, p. 210.
5. Felipe Fernández-Armesto, *Near a Thousand Tables: A History of Food*, pp. 68–69. Also, Jared Diamond, in *Guns, Germs, and Steel*, discusses how keeping livestock, along with people gathering into larger communities, was a recipe for epidemics. He explains that Europeans who weren't strong enough to survive the diseases had already died.
6. Crosby, p. 76. Essig, p. 122.
7. Essig, pp. 125–28.
8. Essig, p. 128.
9. Waverley Root, *Food: An Authoritative and Visual Dictionary of the Foods of the World*, p. 376.
10. Crosby, p. 79.
11. Robin Varnum, *Álvar Núñez Cabeza de Vaca, American Trailblazer*, Norman: University of Oklahoma Press, 2014, p. 193. Essig, p. 127. Crosby, p. 78.
12. Laurens van der Post, *African Cooking*, New York: Time-Life Books, 1970, pp. 113–15.
13. Crosby, p. 79.
14. Lizzie Collingham, *Curry: A Tale of Cooks and Conquerors*, New York: Oxford University Press, 2006, pp. 62, 66–69.
15. David B. Danbom, *Born in the Country: A History of Rural America*, p. 25.
16. Basque fishermen, who appear possibly to have pursued cod as far west as the coastal waters of Canada even before Columbus headed for the Caribbean, are not mentioned, simply because they were not there to settle new lands. They were getting wealthy selling salt cod in Europe.
17. W. W. Buckland and Peter Stein, *A Text-Book of Roman Law: From Augustus to Justinian*, New York: Cambridge University Press, 1963, p. 184. Virginia DeJohn Anderson, *Creatures of Empire: How Domestic Animals Transformed Early America*, p. 78.
18. S. J. Cornelius Michael Buckley, in the Foreword to *Those Terrible Middle Ages: Debunking the Myths*, by Régine Pernoud, San Francisco: Ignatius Press, 2000, p. 7.
19. Charles Darwin, *The Descent of Man*, London: Penguin Books, 2004. p. 183.
20. Standage, *An Edible History of Humanity*, p. 99.

21. "Document: The Purchase of Manhattan Island, 1626," *Dutch New York: Rediscover 400 Years of History*, http://www.thirteen.org/dutchny/interactives/manhattan-island/.

22. "History of Wall Street," Library of Congress Business Research Services, http://www.loc.gov/rr/business/wallstreet/wallstreet.html. Lyall Watson, *The Whole Hog: Exploring the Extraordinary Potential of Pigs*, p. 110. Anderson, pp. 150–51.

23. Roger M. Blench and Kevin C. MacDonald, editors, *The Origins and Development of African Livestock*, http://www.rogerblench.info/Ethnoscience/Animals/Livestock/Pigs%20in%20Africa%20paper.pdf. "History of Slavery and Colonization in South Africa," *South African History Online*, http://www.sahistory.org.za/south-africa-1652-1806/history-slavery-and-early-colonisation-sa.

24. Anderson, p. 97.

25. Suman Roy and Brooke Ali, *From Pemmican to Poutine: A Journey Through Canada's Culinary History*, Toronto, Ontario: Key Publishing House, 2010, pp. 56–60. "Tourtière," The Canadian Encyclopedia. "The Voyageurs," http://digital.library.mcgill.ca/nwc/history/08.htm.

26. The brands of French-Canadian pea soup that I have found most commonly are Habitant and Campbell's.

27. "A Brief History of New Sweden in America," *The Swedish Colonial Society*, http://colonialswedes.net/History/History.html. Dorothy Giles, *Singing Valleys: The Story of Corn*, New York, Random House, 1940, p. 108.

28. Crosby, p. 78. Watson, p. 109.

29. Douglas Taylor, *Anthropological Papers, No. 3, The Caribs of Dominica*, Smithsonian Institution Bureau of American Ethnology; Washington, D.C., United States Government Printing Office, 1938, p. 139. Tannahill, *Food in History*, p. 222. Crosby, p. 76.

30. Wayne Curtis, *And a Bottle of Rum: A History of the New World in Ten Cocktails*, New York: Three Rivers Press, 2007, pp. 24–26. Essig, pp. 128–29.

31. Essig, p. 132.

32. Anderson, pp. 6, 8. Maureen Ogle, *In Meat We Trust: An Unexpected History of Carnivore America*, p. 3.

33. Anderson, p. 45.

34. For a more detailed narrative of the importance of corn, from the first successful colonies up to the present day, you may want to read *Midwest Maize: How Corn Shaped the U.S. Heartland*.

35. Danbom, p. 29.

36. Anderson, pp. 133–40.

37. Anderson, p. 147.

38. Anderson, pp. 101, 110, 147–48. Barry Estabrook, *Pig Tales: An Omnivore's Quest for Sustainable Meat*, p. 52.

39. Kathleen Curtin, Sandra L. Oliver, and Plimouth Plantation, *Giving Thanks: Thanksgiving Recipes and History, from Pilgrims to Pumpkin Pie*, New York: Clarkson Potter Publishers, 2005, p. 103.

40. Danbom, p. 33. Anderson, pp. 162–64. Essig, p. 138.

41. Essig, pp. 132, 136. Anderson, pp. 11. Ogle, p. 6.

42. Anderson, p. 151.

43. Anderson, pp. 9, 17–18, 139, 176–77, 212, 237.

44. Anderson, pp. 100, 303n22.

45. Watson, pp. 121–22.

Chapter 5. American Pig: Shaping American Culture, Agriculture, and Foodways

1. David B. Danbom, *Born in the Country: A History of Rural America*, pp. 55–57.

2. Text of Proclamation can be found at http://www.ushistory.org/us/9a.asp.

3. National Road Museum, Norwich, Ohio.

4. National Road Museum, Norwich, Ohio.

5. The word *corn* means "grain," or, more specifically, "the dominant cereal crop of a region." In England, wheat is corn, which is why maize was sometimes called Virginia wheat. Most, however, simply acknowledged that it was the grain on which Native Americans relied, calling it Indian corn. Cynthia Clampitt, *Midwest Maize: How Corn Shaped the U.S. Heartland*, p. 1.

6. Roger Horowitz, *Putting Meat on the American Table*, pp. 1, 2, 12.

7. Leandra Zim Holland, *Feasting and Fasting with Lewis and Clark: A Food and Social History of the Early 1800s*, Emigrant, Montana: Old Yellowstone Publishing, 2003, p. 97.

8. Mark Essig, *Lesser Beasts*, pp. 149–50.

9. Waverley Root, *Food: An Authoritative and Visual Dictionary of the Foods of the World*, p. 377.

10. Lyall Watson, *The Whole Hog: Exploring the Extraordinary Potential of Pigs*, p. 122.

11. Essig, p. 145. Clampitt, p. 93.

12. Horowitz, pp. 1, 12. Virginia DeJohn Anderson, *Creatures of Empire: How Domestic Animals Transformed Early America*, p. 112.

13. Horowitz, p. 44.

14. Frederick Marryat, *A Diary in America* (1839), quoted in *American Food Writing*, edited by Molly O'Neill, pp. 30, 32.

15. Rachel Laudan, *Cuisine and Empire: Cooking in World History*, pp. 248–49.

16. Horowitz, p. 43.

17. Among the poor, in both rural and urban areas, north and south, salt pork was a mainstay of the diet and often the only available meat. For a more detailed discussion on this, see *Food in the Gilded Age: What Ordinary Americans Ate* by Robert Dirks (Rowman & Littlefield, 2016).

18. Estelle Woods Wilcox, *Buckeye Cookery and Practical Housekeeping*, p. 175.

19. Adrian Miller, *Soul Food: The Surprising History of an American Cuisine, One Plate at a Time*, pp. 92–96.

20. "History of Peanuts and Peanut Butter," National Peanut Board, http://nationalpeanutboard.org/the-facts/history-of-peanuts-peanut-butter/.

21. Cynthia D. Bertelsen, "A Mess of Pottage and a Bite of Bitter Greens: A Brief Look at the Invisible Ethnicity of the English," January 12, 2016. Rachel Lauden, "What Makes This English Cooking?" June 20, 2016, http://www.rachellaudan.com/2016/06/three-cookbooks-that-evoke-memories-of-english-farmhouse-cuisine.html.

22. Miller, p. 45.

23. Stephen Schmidt, "When Did Southern Begin?" *Manuscript Cookbooks Survey*. November 2015. http://www.manuscriptcookbookssurvey.com/when-did-southern-begin/. In this article, the author goes so far as to attribute the concept of clearly defined regional cooking to cookbook publishers in the mid-1900s.

24. "Invention," Can Manufacturers Institute, http://www.cancentral.com/content/nicolas-appert-father-canning; "The First U.S. Can Opener," *Connecticut History*, http://connecticuthistory.org/the-first-us-can-opener-today-in-history.

25. Peter Smith, "Underwood's Deviled Ham: The Oldest Trademark Still in Use," *Smithsonian*, March 9, 2012, http://www.smithsonianmag.com/arts-culture/underwoods-deviled-ham-the-oldest-trademark-still-in-use-119136583/.

26. Clampitt, pp. 81–87.

27. Catherine McNeur, *Taming Manhattan: Environmental Battles in the Antebellum City*, Cambridge, MA: Harvard University Press, 2014, p. 2.

28. Essig, pp. 181, 184.

29. McNeur, pp. 136–37, 161.

30. Maureen Ogle, *In Meat We Trust: An Unexpected History of Carnivore America*, p. 14.

31. Essig, pp. 146, 161.

32. Clampitt, p. 99.

33. Ogle, pp. 26–28.

34. Christian Wolmar, *The Great Railroad Revolution: The History of Trains in America*, p. 215.

Chapter 6. Corn Belt/Hog Belt: How Hogs and Hominy Helped Define a Region

1. James R. Shortridge, *The Middle West*, pp. 20–21.

2. A more detailed discussion of how and when the terms arose, and what precisely they described, can be found in chapter 3, "Birth of the Midwest and the Corn Belt," in Cynthia Clampitt, *Midwest Maize: How Corn Shaped the U.S. Heartland*.

3. U.S. Department of Agriculture, Census of Agriculture, http://www.agcensus.usda.gov/Publications/2012/Online_Resources/Highlights/Hog_and_Pig_Farming/.

4. There were some devastating conflicts, though some of the best known were later and farther west. It wasn't until 1830, when Andrew Jackson proposed the

Indian Removal Act, that the government's position changed. Native Americans were paid to move west—whether they wanted to go or not. This act mainly targeted states in the southeast, but it reinforced the idea for settlers that the land was for sale.

5. Robert C. Pollock, "Emerson," from *Doctrine and Experience: Essays in American Philosophy*, Vincent G. Potter ed., New York: Fordham University Press, 1988, p. 67.

6. V. S. Robert Jennings, *Sheep, Swine, and Poultry: Embracing the History and Varieties of Each; Their Feeding and Management, etc.*, Philadelphia: John E. Potter and Company, 1864, p. 255.

7. Maureen Ogle, *In Meat We Trust: An Unexpected History of Carnivore America*, p. 11.

8. Peter Kaminsky, *Pig Perfect: Encounters with Remarkable Swine and Some Great Ways to Cook Them*, pp. 158–59.

9. Lyall Watson, *The Whole Hog: Exploring the Extraordinary Potential of Pigs*, p. 125.

10. Allan G. Bogue, *From Prairie to Corn Belt: Farming on the Illinois and Iowa Prairies in the Nineteenth Century*, p. 109.

11. John C. Hudson, *Making the Corn Belt*, pp. 66–68, 72, 76. Mark Essig, *Lesser Beasts: A Snout-to-Tail History of the Humble Pig*, p. 159–61.

12. Watson, pp. 125–26. Essig, pp. 160–61. Hudson, p. 84–85.

13. Hudson, pp. 68–72. William Cronon, *Nature's Metropolis: Chicago and the Great West*, pp. 222–23.

14. Jay L. Lush and A. L. Anderson, "The Genetic History of Poland-China Swine," *Journal of Heredity*, vol. 30 no. 4, April 1939, http://jhered.oxfordjournals.org/content/30/4/149.extract.

15. Crude oil that bubbled to the surface had certainly been noticed previously, but without any impact. The first place it was clearly both observed and used was in Pennsylvania, where early Native Americans skimmed oil off the surface of a stream and used it for religious rituals. The area where they collected the crude oil was named Oil Creek by settlers in 1755. Initial use by Europeans was as medicine. The actual discovery of petroleum and its usefulness was in the 1850s. It was initially used as a lubricant and, as kerosene, in lamps, with gasoline being considered a waste product. Brian Black, "Petroleum History, United States," *The Encyclopedia of the Earth*, http://www.eoearth.org/view/article/155205/.

16. Essig, pp. 208–09. Kaminsky, p. 73.

17. "French Detroit 1701–1760," *Detroit History*, http://historydetroit.com/eras/french_rule_1701-1760.php.

18. "A Brief History of St. Louis," https://www.stlouis-mo.gov/visit-play/stlouis-history.cfm.

19. Hudson, p. 81. Essig, p. 169.

20. Bogue, p. 104.

21. Mark Kurlansky, *Salt: A World History*, pp. 251–55.

22. Hudson, p. 82.

23. Quoted in Cronon, p. 229.

24. Ogle, p. 15. Cronon, p. 229.

25. "A History of Innovation: Our Humble Beginnings," Procter & Gamble, http://us.pg.com/who_we_are/heritage/history_of_innovation. Eric Jay Dolin, "From Whale Oil and Beyond," *Boston.com News*, October 11, 2007. Hudson, p. 85.

26. Ogle, pp. 10–11. "Our Heritage," Anheuser-Busch. http://anheuser-busch.com/index.php/our-heritage/history/.

27. Bogue, p. 103.

28. Watson, p. 126. Cronon, p. 230.

29. Amy Murrell Taylor, "The Border States," National Park Service, https://www.nps.gov/resources/story.htm%3Fid%3D205.

30. Jon K. Lauck, *The Lost Region: Toward a Revival of Midwestern History*, pp. 18–19. The Civil War Home Page, http://www.civil-war.net/pages/troops_furnished_losses.html.

31. Andrew Smith, *Starving the South: How the North Won the Civil War*, pp. 74–76.

32. Mark Essig, "The Great Appalachian Hog Drives," *Atlas Obscura*, May 4, 2015, http://www.atlasobscura.com/articles/the-great-appalachian-hog-drives.

33. Smith, pp. 20–23. McIlhenny Company, http://www.tabasco.com/mcilhenny-company/about/. Kurlansky, pp. 257–75.

34. Smith, pp. 79, 80, 87.

35. "William T. Sherman," Civil War Trust, http://www.civilwar.org/education/history/biographies/william-t-sherman.html. Smith, pp. 177–81.

36. "Was the Homestead Act Color Blind?" NebraskaStudies.org.

37. R. Douglas Hurt, *American Agriculture*, pp. 192–94.

38. Morrill Act, Library of Congress, http://www.loc.gov/rr/program/bib/ourdocs/Morrill.html.

39. "Morrill Acts," Law and Higher Education, http://lawhigheredu.com/90-morrill-acts.html.

40. Trina R. Williams Shanks, "The Homestead Act of the Nineteenth Century and its Influence on Rural Lands," Center for Social Development, p. 6.

41. http://www.nal.usda.gov/lincolns-milwaukee-speech.

42. "The Pacific Railway," http://railroad.lindahall.org/essays/brief-history.html.

43. "USDA History" and "Abraham Lincoln an Agriculture," U.S. Department of Agriculture, National Agricultural Library, http://www.nal.usda.gov/history-art-and-biography/usda-history.

44. Dominic A. Pacyga, *Slaughterhouse: Chicago's Union Stock Yard and the World It Made*, p. 29.

45. "Meatpacking Technology," *Slaughterhouse to the World*, http://www.chicagohs.org/history/stockyard/stock2.html.

46. Pacyga, pp. 1, 39.

47. "Meatpacking," Encyclopedia of Chicago, http://www.encyclopedia.chicagohistory.org/pages/804.html. Pacyga, p. 3.

48. Cronon, p. 232.

49. Cronon, p. 233.

50. Pacyga, pp. 45–58. "The Birth of the Chicago Stock Yards," *Slaughterhouse to the World*, Chicago Historical Society, http://www.chicagohs.org/history/stockyard/stock1.html.

51. Essig, p. 199. Ogle, pp. 70–73.

52. "Dr. John Harvey Kellogg—Inventor of Kellogg's Corn Flakes," University of Texas Health Science Center, http://library.uthscsa.edu/2014/05/dr-john-harvey-kellogg-inventor-of-kelloggs-corn-flakes/. Essig, p. 199. Rae Katherine Eighmey, *Food Will Win the War: Minnesota Crops, Cooks, and Conservation During World War I*, St. Paul, MN: Minnesota Historical Society Press, 2010, p. viii. "How to Cook Salt Pork," distributed by the U.S. Department of Agriculture in 1933, https://archive.org/stream/CAT10679515/CAT10679515_djvu.txt.

53. Cronon, pp. 257–59. "Death of the Stock Yards," *Slaughterhouse to the World*, http://www.chicagohs.org/history/stockyard/stock9.html.

54. Bogue, p. 103.

55. "The Rise and Fall of the Omaha Stockyards," http://www.livinghistoryfarm.org/farminginthe50s/money_14.html.

56. "The Roots of Kohler," http://www.us.kohler.com/us/The-Kohler-Heritage/content/CNT900083.htm.

57. W. W. Burns, "The Hog," in *The Iowa Yearbook of Agriculture*, vol. 3, Des Moines, IA: B. Murphy, State Printer, 1903, p. 42.

58. Jolene Stevens, *Pigs! Lifting Mortgages, People, and Communities: A History of the Pork Industry in Lyon, Plymouth, and Sioux Counties in Iowa*, p. 4.

59. Watson, p. 126.

Chapter 7. Versatile Pig: The Parts and How We Use Them

1. "The Hog," chapter 77 of Book VIII in Pliny the Elder's *Natural History*, p. 344 of the 1855 John Bostock translation.

2. Mrs. *Beeton's Book of Household Management*, 1861, reprinted by Oxford University Press, 2000, p. 188.

3. "Fresh Ham/Fresh Leg," National Pork Board, http://www.porkbeinspired.com/cuts/fresh-hamfresh-leg/.

4. Mark Essig, *Lesser Beasts: A Snout-to-Tail History of the Humble Pig*, p. 201. Alan Davidson, *The Oxford Companion to Food*, pp. 368–69. "Smithfield Ham," *Cook's Info*, http://www.cooksinfo.com/smithfield-ham.

5. Roger Horowitz, *Putting Meat on the American Table: Taste, Technology, Transformation*, pp. 45–46.

6. Essig, p. 201.

7. "Smithfield Marketplace Guide on How to Prepare a Country Ham," https://www.smithfieldmarketplace.com/about-country-hams.

8. Mower County Historical Society, Austin, MN. Hormel Historic Home, Austin, MN. And information supplied by Hormel Foods.

9. Horowitz, p. 62.

10. Davidson, p. 47. Ari Weinzweig, *Zingerman's Guide to Better Bacon*, p. 22.

11. Essig, pp. 201–02. Horowitz, p. 63.

12. Horowitz, pp. 55, 63.

13. Bruce Kraig, *Hot Dog: A Global History*, p. 65.

14. Horowitz, pp. 63–64.

15. Horowitz, pp. 65–69.

16. Monica Davey, "Trade in Pork Bellies Comes to an End, but the Lore Lives," *New York Times*, July 30, 2011.

17. Zosia Chustecka, "WHO Clarifies Processed Meat/Cancer Link After 'Bacon-gate,'" November 2, 2015, *Medscape Medical News*, http://www.medscape.com/viewarticle/853566.

18. Martijn B. Katan, "Nitrate in Foods: Harmful or Healthy," *The American Journal of Clinical Nutrition*, May 20, 2009. Caroline Praderioa, "Nitrites & Nitrates: Are They Harmful or Actually Healthful?" *Prevention Magazine*, January 20, 2015.

19. Horowitz, p. 46.

20. Robert Dirks, *Come & Get It!: Mcdonaldization and the Disappearance of Local Food From a Central Illinois Community*, p. 56.

21. H. T. Fredeen and J. A. Newman, "Rib and Vertebral Numbers in Swine: Variations Observed in Large Population," *Canadian Journal of Animal Science*, vol. 42 no. 2, December 1962, pp. 232–39.

22. Dann Woellert, "Frugal Germans in Porkopolis and Our Throwaway Cuts of Meat," *The Food Etymologist*, October 22, 2015.

23. *Merriam-Webster's Dictionary*.

24. Dirks, p. 61.

25. LeRoi Jones, "Soul Food," *Home: Social Essays*, 1966, quoted in *American Food Writing*, edited by Molly O'Neill, p. 385.

26. "Fire in the Hole: Barbecue and American Culture," *The Economist*, December 16, 2010.

27. Katherine Sacks with Will Blunt, "The Product: Porchetta, Digging into Italian Culinary History," *Starchefs*, July 2012.

28. John T. Edge, "As Tasty Morsels, Pig Wings Take Flight," *New York Times*, November 29, 2011.

29. Originally published in the *Baltimore Evening Sun*, November 4, 1928; quoted in *American Food Writing*, edited by Molly O'Neill.

30. Michael Ruhlman and Brian Polcyn, *Charcuterie: The Craft of Salting, Smoking, and Curing*, New York: W. W. Norton and Company, 2005, pp. 100–02. Sharon Tyler Herbst, *Food Lover's Companion*, p. 545.

31. Horowitz, pp. 92–95. Viskase® History, http://viskase.com/about-us/viskase-history/.

32. Davidson, p. 443.

33. Herbst, p. 107.

34. Essig, pp. 209–11. Brian Black, "Petroleum History, United States," *The Encyclopedia of the Earth*, http://www.eoearth.org/view/article/155205/.

35. Julie R. Thomson, "10 Reasons You Should Be Cooking with Lard," *Huffington Post*, April 28, 1014, http://www.huffingtonpost.com/2014/04/28/cooking-with-lard-baking_n_5212804.html.

36. Woellert, https://dannwoellertthefoodetymologist.wordpress.com/2015/10/22/frugal-germans-in-porkopolis-and-our-throwaway-cuts-of-meat/.

37. Essig, p. 171. Daniel W. Gade, "Hogs (Pigs)," *The Cambridge World History of Food*, p. 541.

38. From *Almanach des Gourmands*, translated from the French by Mark Kurlansky, in *Choice Cuts*, London: Penguin Books, 2004, p. 237.

39. Allan G. Bogue, *From Prairie to Corn Belt*, p. 106–08.

40. D. Phillip Sponenberg, Jeannette Beranger, and Alison Martin, *Introduction to Heritage Breeds: Saving and Raising Rare-Breed Livestock and Poultry*, pp. 220–21. National Swine Registry: Duroc, http://nationalswine.com/about/about_breeds/duroc.php. These three are not the only breeds developed in the United States; they are just the ones that are currently in the spotlight, along with the British breeds that remain popular. In the 1800s, the Poland China was one of the U.S.-developed stars, and it is still around. Some breeds simply developed/were bred into breeds that are widely raised today but too common to be heritage breeds. Many breeds from the 1800s vanished when they declined in popularity and people stopped raising them. This happened to most of the lard-type pigs. The Mulefoot is one of only three lard-type pigs that remain today, the others being the Choctaw and the Guinea Hog. For more, check out "The History of Heritage Pigs" on the Livestock Conservancy site, http://livestockconservancy.org/.

41. Melissa Anders, "Michigan Farmers Raise Furry, Hungarian, Mangalitsa Pigs to Produce the 'Kobe Beef' of Pork," *MLive Michigan*, August 15, 2013.

Chapter 8. American Icons: Barbecue, Hot Dogs, and SPAM®

1. James Villas, *American Taste: A Celebration of Gastronomy Coast to Coast*, p. 42. "Early Carolina Settlement: Barbados Influence," Lowcountry Digital History Initiative, http://ldhi.library.cofc.edu/exhibits/show/africanpassageslowcountryadapt/sectionii_introduction/barbados_influence.

2. Sylvia Lovegren, "Barbecue," *The Oxford Companion to American Food and Drink*, pp. 35–36.

3. *Founders Online*, Diary Entry, May 27, 1769, http://founders.archives.gov/documents/Washington/01-02-02-0004-0013-0027.

4. Adrian Miller, "Toward a Theology of Barbecue," *Faith and Leadership*, https://www.faithandleadership.com/adrian-miller-toward-theology-barbecue.

5. Lovegren, pp. 35–36.

6. Miller, "Theology of Barbecue."

7. Horowitz, p. 80. Bruce Kraig, *Hot Dog: A Global History*, pp. 16–22.

8. Kraig, p. 23.

9. Kraig, pp. 12, 43.

10. Horowitz, p. 79.

11. Bruce Kraig and Patty Carroll, *Hot Dog Culture in America: Man Bites Dog*, Lanham, MD: AltaMira Press, 2012, pp. 51–52.

12. William Grimes, *Eating Your Words*, New York: Oxford University Press, 2004, p. 165.

13. Horowitz, pp. 82–84.

14. Kraig, pp. 84–87. Horowitz, pp. 86–87.

15. Horowitz, pp. 75, 76, 91, 95, 99. Kraig, pp. 63–69. Andrew F. Smith, *Eating History: Thirty Turning Points in the Making of American Cuisine*, p. 287. Iowa Business Hall of Fame, http://www.iowabusinesshalloffame.com/inductees/townsend-ray.html.

16. Erin DeJesus, "A Brief History of Spam, an American Meat Icon," *Eater*, July 9, 2014, http://www.eater.com/2014/7/9/6191681/a-brief-history-of-spam-an-american-meat-icon. Ed Grabianowski, "How Spam Works," *How Stuff Works*, October 3, 2007, http://science.howstuffworks.com/innovation/edible-innovations/spam-food2.htm. Carolyn Wyman, "Spam," *The Oxford Encyclopedia of Food and Drink in America*, Volume 2, pp. 342–43. Additional input from Hormel Foods.

Chapter 9. Local Pig: Influences and Specialties in the Heartland

1. Jon K. Lauck, *The Lost Region: Toward a Revival of Midwestern History*, p. 24.

2. A list of cookbooks from the mid- to late 1800s that I used for research can be found in the "Historic Cookbooks" section at the end of the book.

3. Brian W. Beltman, "Ethnicity," *The Greenwood Encyclopedia of American Regional Cultures: The Midwest*, p. 130.

4. John C. Hudson, *Making the Corn Belt: A Geographical History of Middle-Western Agriculture*, p. 63.

5. Roger Biles, *Illinois: A History of the Land and Its People*, DeKalb, IL: Northern Illinois University Press, 2005, p. 56.

6. Beltman, p. 130.

7. Biles, p. 60

8. Beltman, p. 131.

9. Harvard University Library, "Immigration to the United States," http://ocp.hul.harvard.edu/immigration/scandinavian.html.

10. Beltman, pp. 136–37.

11. John Radzilowski, "Poles," *Encyclopedia of the Great Plains*.

12. Lucy M. Long, "Food," *The Greenwood Encyclopedia of American Regional Cultures: The Midwest*, p. 310.

13. "Chinatown," *Encyclopedia of Chicago*, http://www.encyclopedia.chicagohistory.org/pages/284.html.

14. Larry O'Dell, "All-Black Towns," *Encyclopedia of the Great Plains*.

15. Beltman, p. 138.

16. Renee M. Laegreid, "Italians," *Encyclopedia of the Great Plains*.

17. Bruce M. Garver, "Czechs," *Encyclopedia of the Great Plains*.

18. "Latinos in Michigan," *Los Repatriados*, University of Michigan, http://www.umich.edu/~ac213/student_projects07/repatriados/history/mihistory.html. Bruce Corrie, "People of Mexican Origin in Minnesota: Myth vs. Reality," *Twin Cities Daily Planet*, April 18, 2008.

19. Long, pp 296–300. Anne Mendelson, *Stand Facing the Stove: The Story of the Women Who Gave America the Joy of Cooking*, New York: Scribner, 2003, pp. 9–14.

20. Dirks, p. 37.

21. "Recipes from the Iron Range," Minnesota Public Radio, May 21, 2006, http://www.mprnews.org/story/2006/05/16/rangerecipes.

22. Adrian Miller, *Soul Food: The Surprising History of an American Cuisine, One Plate at a Time*, pp. 8–10, 15, 101.

23. Claire Suddath, "A Brief History of Barbecue," *Time*, July 3, 2009.

24. Iowa Pork Producers Association.

Chapter 10. Transformed Pig: Recipes for Specialties

1. *Lost Iron Range*, documentary, WDSE, WRPT-PBS, 2014.

2. Tim Dondero, "Soups in Ancient Rome," *Online Athens: Athens Banner-Herald*, April 18, 2010, http://onlineathens.com/stories/041810/liv_610999089.shtml.

3. Estelle Woods Wilcox, *Buckeye Cookery and Practical Housekeeping*, p. 174.

Chapter 11. Popular Pig: The Rise of Pig Obsession

1. U.S. Department of Agriculture National Agricultural Library, http://www.nal.usda.gov/lincolns-milwaukee-speech.

2. Betty Harper Fussell, *The Story of Corn*, Albuquerque: University of New Mexico Press, 1992, p. 313.

3. Christy Campbell, *Eat & Explore Ohio: Cookbook and Travel Guide*, Lena, MS: Great American Publishers, 2015, p. 165. http://www.porkfestival.org/schedule/.

4. http://www.tiptoncountyporkfestival.com/.

5. http://www.chmoorehomestead.org/apple-pork.htm.

6. "Gary and Carolyn Wells," Barbecue Hall of Fame Legacy Inductees, http://www.barbecuehalloffame.com/p/getconnected/legacy-inductees/237. And from Carolyn Wells.

7. Taste of Elegance: http://www.mopork.com/events/taste-of-elegance/. Cochon 555: http://cochon555.com/mission/.

Chapter 12. Farm to Table Pig: People Who Raise Pigs and Create Pork

1. This report from the Center for Disease Control offers more details on the "Chinese footprint": http://wwwnc.cdc.gov/eid/article/20/5/14-0195_article. This report, from the National Pork Board, offers more info on the virus but also empha-

sizes that pork is completely safe for consumption: http://www.pork.org/national-pork-board-statement-porcine-epidemic-diarrhea-virus-pedv/.

2. For a related dish and a bit more history, see the recipe for Goetta in chapter 10.

Chapter 13. Cherished Pig: Traditions That Have Changed and Some Worth Keeping

1. Evan D. G. Fraser and Andrew Rimas, *Empires of Food: Feast, Famine, and the Rise and Fall of Civilizations*, pp. 43, 47, 52–57. Neville Morley, "The Profits of Empire," in the *Cambridge Illustrated History of the Roman World*, p. 295.

2. U.S. Department of Agriculture, "Briefing on the Status of Rural America," http://www.usda.gov/documents/Briefing_on_the_Status_of_Rural_America_Low_Res_Cover_update_map.pdf.

3. "Historical Timeline: Farmers and the Land," *Growing a Nation: The Story of American Agriculture*, https://www.agclassroom.org/gan/timeline/farmers_land.htm. *The CIA World Factbook: United States, Labor force-by occupation*, https://www.cia.gov/library/publications/the-world-factbook/geos/us.html.

4. For more information on the Animal Welfare Approved program, you can visit their site: http://animalwelfareapproved.org/.

5. Julian Wiseman, *The Pig: A British History*, pp. 20–59. Allen G. Bogue, *From Prairie to Corn Belt*, p. 106.

6. Dr. Steve Henry, veterinary clinician specializing in swine and population medicine.

7. http://www.agcensus.usda.gov/Publications/2012/Online_Resources/Highlights/Hog_and_Pig_Farming/.

8. Nichola Fletcher, *Charlemagne's Tablecloth: A Piquant History of Feasting*, pp. 125–26.

9. The "purple stuff" of Alice's memory would have been saltpeter or pink curing salt, important for the curing process and for preserving the color of meat.

10. Mark Kurlansky, *Salt: A World History*, p. 95. https://www.prosciuttodiparma.com/en_UK/prosciutto/pigs.

11. Jan Fialkow, "And the Banned Played On," *Gourmet Retailer*, February/March 2014, http://www.gourmetretailer.com/article-and_the_banned_played_on-6822.html.

12. http://www.4-h.org/about/.

13. https://www.ffa.org/about/what-is-ffa.

Chapter 14. Problems and Promises: Issues, Concerns, Discussions, and Hopes

1. Matias B. Vanotti, Ariel A. Szogi, Patrick G. Hunt, Patricia D. Millner, and Frank J. Humenik, "Development of environmentally superior treatment system to replace anaerobic swing lagoons in the USA," *Science Direct*, August 21, 2006, http://pubag.nal.usda.gov/pubag/downloadPDF.xhtml?id=9314&content=PDF.

2. Karen Uhlenhuth, "Missouri project 'a new level' in manure biogas production," *Midwest Energy News*, April 7, 2016, http://midwestenergynews.com/2016/04/07/missouri-project-a-new-level-in-manure-biogas-production/. A. E. Roeslein, "Project to convert manure into energy celebrates milestone," *RAE News*, September 14, 2016, http://roesleinalternativeenergy.com/project-to-convert-manure-into-energy-celebrates-milestone/.

3. http://www.pork.org/production-topics/environmental-sustainability-efforts-pork-production/carbon-footprint-pork-production-calculator/.

4. Mark Essig, *Lesser Beasts: A Snout-to-Tail History of the Humble Pig*, pp. 203–04.

5. "Vesicular Exanthema," *The Pig Site*, http://www.thepigsite.com/diseaseinfo/129/vesicular-exanthema/. U.S. Department of Agriculture, "Vesicular Exanthema Eradication," November 1955, https://archive.org/stream/vesicularexanthe19unit/vesicularexanthe19unit_djvu.txt.

6. Essig, p. 204.

7. Kenneth J. Carpenter, "The Nobel Prize and the Discovery of Vitamins," June 22, 2004, http://www.nobelprize.org/nobel_prizes/themes/medicine/carpenter/. "Vitamin B_{12}," "Medical Discoveries," http://www.discoveriesinmedicine.com/To-Z/Vitamin-B12.html. Essig, p. 212.

8. Dan Charles, "Milloins of Chickens to be Killed as Bird Flu Outbreak Puzzles Industry," NPR/All Things Considered, April 21, 2015, http://www.npr.org/sections/thesalt/2015/04/21/401319019/5-million-chickens-to-be-killed-as-bird-flu-outbreak-puzzles-industry.

9. John M. Barry, *The Great Influenza: The Epic Story of the Deadliest Plague in History*, New York: Penguin Books, 2004, pp. 112–13, 177–78, 446. Ian H. Brown, "The Epidemiology and Evolution of Influenza Viruses in Pigs," *Veterinary Microbiology*, vol. 74 no. 1–2, May 22, 2000, pp. 29–46. "Influenza," Medical Ecology, http://www.medicalecology.org/diseases/influenza/influenza.htm#sect2.1.

10. Meredith Davis and Theopolis Waters, "Killer virus spreads unchecked through U.S. hog belt, pushing pork to record," *Reuters*, April 27, 2014, http://www.reuters.com/article/us-usa-pork-pig-virus-idUSBREA3Q0IC20140427.

11. Jan Fialkow, "And the Banned Played On," *Gourmet Retailer*, February/March 2014, http://www.gourmetretailer.com/article-and_the_banned_played_on-6822.html.

12. Daniel W. Gade, "Hogs (Pigs)," *The Cambridge World History of Food*, p. 541. Joseph B. Treaster, "Haiti Replenishing a Most Valuable Asset: The Pig," *The New York Times*, October 3, 1984. For more info on ASF, see "African Swine Fever: How can Global Spread be Prevented?" by Solenne Costard, Barbara Wieland, William de Glanville, Ferran Jori, Rebecca Rowlands, Wilna Vosloo, François Roger, Dirk U. Pfeiffer, and Linda K. Dixon, available through the U.S. National Library of Medicine, National Institutes of Health, http://www.ncbi.nlm.nih.gov/pmc/articles/PMC2865084/.

13. http://www.wral.com/news/local/asset_gallery/5225107/.

14. Stephen Schmidt, "CAFNR geneticist teams up with Kansas State researcher to make PRRS-resistant pigs," *CAFNR News*, December 8, 2015, http://cafnrnews.com/2015/12/big-cat-collaboration/.

15. "The Market Works: Highlighting Progress in the Meat Industry," http://www.themarketworks.org/stats.

16. *Wild Hogs in Michigan: Threats to Property, Detection, and Control*, Michigan Wildlife Conservancy, Bengel Wildlife Center, Bath, Michigan. Amy Nordrum, "Can Wild Pigs Ravaging the U.S. Be Stopped?" *Scientific American*, October 21, 2014.

17. Barry Estabrook, *Pig Tales: An Omnivore's Quest for Sustainable Meat*, pp. 47–53. John Morthland, "A Plague of Pigs in Texas," *Smithsonian.com*, January 2011, p. 1, http://www.smithsonianmag.com/science-nature/a-plague-of-pigs-in-texas-73769069/?no-ist.

18. Morthland, p. 1. "Threats to Native Wildlife," *Wild Pig Info*, Mississippi State University Extension, http://wildpiginfo.msstate.edu/threats-wildlife-wild-pigs.html.

19. Morthland, p. 1.

20. Julie Turner-Crawford, "Feral Hogs Can Impact Livestock," *Ozarks Farm and Neighbor*, http://www.ozarksfn.com/ark-articles-aamp-stories-editorial-92/94-ag-visors-arkansas/7973-feral-hogs-can-impact-livestock-.html.

21. Ryan Schlehuber, "WOODS and WATER: On the Cusp of a Feral Swine Outbreak?" *The Daily News*, September 30, 2014, http://thedailynews.cc/2014/09/30/woods-water-on-the-cusp-of-a-feral-swine-outbreak/.

22. Gail Keirn, "We Can't Barbecue Our Way Out: Why Feral Swine Management Requires a National Approach," U.S. Department of Agriculture, April 4, 2014, http://blogs.usda.gov/2014/04/04/we-cant-barbecue-our-way-out-why-feral-swine-management-requires-a-national-approach/.

23. "Kansas Battle to Reduce Feral Hog Population Seems to be Working," *The Wichita Eagle*, March 9, 2014, http://www.kansas.com/sports/outdoors/article1136670.html.

24. "Human Population: Population Growth," *Population Reference Bureau*, http://www.prb.org/Publications/Lesson-Plans/HumanPopulation/PopulationGrowth.aspx; U.S. Census Bureau Population Clock, https://www.census.gov/popclock/.

25. "World Hunger Falls to Under 800 Million, Eradication Is Next Goal," World Food Programme, https://www.wfp.org/news/news-release/world-hunger-falls-under-800-million-eradication-next-goal-0.

26. According to the U.S. Department of Agriculture Census of Agriculture, 83 percent of the farms that raise pigs are owned by families or individuals, and another 7 percent are partnerships. http://www.agcensus.usda.gov/Publications/2012/Online_Resources/Highlights/Hog_and_Pig_Farming/.

27. Loren Grush, "Why pigs are so valuable for medical research," *Fox News Health*, May 9, 2014, http://www.foxnews.com/health/2014/05/09/why-pigs-are-so-valuable-for-medical-research.html.

28. Sam Kean, "The Strange, Gruesome Search for Substance X," *Distillations Magazine*, vol. 1 no. 1, Spring 2015. "Role of Endorphins Discovered, 1975," http://www.pbs.org/wgbh/aso/databank/entries/dh75en.html.

29. On January 26, 2017, it was reported in *Cell* magazine that the first human cells had been grown in a pig embryo. It was not a useful amount, as pigs develop at a

much faster rate than humans, but it was the first pig-human "chimera," as creatures with two animals' cells are called.

30. "Outlook for Major Pork Importing Countries," U.S. Meat Export Federation, 2016, https://www.usmef.org/news-statistics/press-releases/outlook-for-major-pork-importing-countries-14403/.

31. Gade, p. 541. U.S. Department of Agriculture Economic Research Service, http://www.ers.usda.gov/topics/animal-products/hogs-pork/trade.aspx. U.S. Meat Export Federation, http://www.usmef.org/news-statistics/press-releases/outlook-for-major-pork-importing-countries-14403/.

32. "Russia's Pork Import Ban: What's the Global Impact?" *National Hog Farmer*, August 18, 2014, http://nationalhogfarmer.com/business/russia-s-pork-import-ban-what-s-global-impact.

33. Gary Schnitkey, "Chinese and U.S. Pork Consumption and Production," *FarmsdocDaily*, University of Illinois, June 25, 2013, http://farmdocdaily.illinois.edu/2013/06/chinese-us-pork-consumption-production.html.

34. Adam Jourdan, "Overcrowding on farms behind mystery of China's floating pigs," *Reuters*, April 24, 2013, http://www.reuters.com/article/us-china-farming-pigs-idUSBRE93N1C720130424.

35. Denny Thomas and Olivia Oran, "China's Appetite for Pork Spurs $4.7 billion Smithfield Deal," *Reuters*, May 29, 2013, http://www.reuters.com/article/us-shuanghui-idUSBRE94S0K920130529.

36. "Smithfield says US pork exports to China jump 45 Percent in First Six Months of Year," *Reuters*, August. 25, 2015, http://www.reuters.com/article/smithfield-foods-pork-china-idUSL1N1101NC20150825.

37. Even Rogers Pay, "What American Hog Farmers Think of the Smithfield Deal," *Modern Farmer*, June 6, 2013. David Pitt, "Pollution concerns heighten hog conflicts," *The Des Moines Register*, February 16, 2015. Lynn Hicks, "Feeding China: High-tech hog farms," *The Des Moines Register*, October 12, 2104.

38. Christopher L. Delgado, "Rising Consumption of Meat and Milk in Developing Countries has Created a New Food Revolution," *The Journal of Nutrition*, vol. 133 no. 11, November 1, 2003.

Bibliography and Sources

These are sources from which substantial information or insight was gained. Books, journal articles, and other resources that supplied a single fact or statistic are referenced only in endnotes.

Bibliography

Albarella, Umberto, Keith Dobney, Anton Ervynck, and Peter Rowley-Conwy, editors. *Pigs and Humans: 10,000 Years of Interaction*. Oxford, UK: Oxford University Press, 2007.

Albarella, Umberto, Keith Dobney, and Peter Rowley-Conwy. "Size and Shape of the Eurasian Wild Boar (*Sus scrofa*), with a View to the Reconstruction of its Holocene History." *Environmental Archaeology*, vol. 14 no. 2, 2009.

Anderson, J. L. *Industrializing the Corn Belt: Agriculture, Technology, and Environment, 1945-1972*. DeKalb: Northern Illinois University Press, 2009.

Anderson, Virginia DeJohn. *Creatures of Empire: How Domestic Animals Transformed Early America*. New York: Oxford University Press, 2004.

Bogue, Allan G. *From Prairie to Corn Belt: Farming on the Illinois and Iowa Prairies in the Nineteenth Century* (second edition). Lanham, MD: Ivan R. Dee Publisher, 2011.

Carcopino, Jérôme. *Daily Life in Ancient Rome*. New Haven, CT: Yale University Press, 1969.

Clampitt, Cynthia. *Midwest Maize: How Corn Shaped the U.S. Heartland*. Champaign: University of Illinois Press, 2015.

Conkin, Paul K. *A Revolution Down on the Farm: The Transformation of American Agriculture Since 1929*. Lexington: The University Press of Kentucky, 2009.

Cosman, Madeleine Pelner. *Fabulous Feasts: Medieval Cookery and Ceremony*. New York: George Braziller, Inc., 1976.

Cronon, William. *Nature's Metropolis: Chicago and the Great West*. New York: W. W. Norton & Company, Inc., 1991.

Crosby, Alfred W., Jr. *The Columbian Exchange: Biological and Cultural Consequences of 1492*. Westport, CT: Praeger Publishers, 2003.

Danbom, David B. *Born in the Country: A History of Rural America* (second edition). Baltimore, MD: The Johns Hopkins University Press, 2006.

Danforth, Randi, Peter Feierabend, and Gary Chassman, editors. *Culinaria: The United States*. New York: Könemann Publishers USA, 1998.

Davidson, Alan. *The Oxford Companion to Food*. New York: Oxford University Press, 1999.

Delgado, Christopher L. "Rising Consumption of Meat and Milk in Developing Countries Has Created a New Food Revolution." *The Journal of Nutrition*, vol. 133 no. 11, November 1, 2003.

Diamond, Jared. *Guns, Germs, and Steel: The Fates of Human Societies*. New York: W. W. Norton and Company, Inc., 1999.

Dirks, Robert. *Come & Get It!: Mcdonaldization and the Disappearance of Local Food From a Central Illinois Community*. Bloomington, IL: McLean County Museum of History, 2011.

Ebrey, Patricia Buckley. *The Cambridge Illustrated History of China*. London, UK: Cambridge University Press, 1996.

Ellis, Peter Berresford. *The Celts: A History*. New York: Carrol and Graff Publishers, 2004.

Essig, Mark. *Lesser Beasts: A Snout-to-Tail History of the Humble Pig*. New York: Basic Books, 2015.

Estabrook, Barry. *Pig Tales: An Omnivore's Quest for Sustainable Meat*. New York: W. W. Norton and Company, 2015.

Faas, Patrick, translated from the Dutch by Shaun Whiteside. *Around the Roman Table: Food and Feasting in Ancient Rome*. New York: Palgrave Macmillan, 2003.

Fernández-Armesto, Felipe. *Near a Thousand Tables: A History of Food*. New York: The Free Press, 2002.

Fletcher, Nichola. *Charlemagne's Tablecloth: A Piquant History of Feasting*. New York: St. Martin's Press, 2004.

Fraser, Evan D. G., and Andrew Rimas. *Empires of Food: Feast, Famine, and the Rise and Fall of Civilizations*. New York: Free Press, 2010.

Freedman, Paul, ed. *Food: The History of Taste*. Berkeley: University of California Press, 2007.

Gade, Daniel W. "Hogs (Pigs)," in *The Cambridge World History of Food*. Cambridge, UK: Cambridge University Press, 2000.

Garavini, Daniela, ed. *Pigs and Pork: History, Folklore, and Ancient Recipes*. Cologne, Germany: Könemann, 1999.

Gies, Frances, and Gies, Joseph. *Daily Life in Medieval Times: A Vivid, Detailed Account of Birth, Marriage and Death; Food, Clothing and Housing, Love and Labor in the Middle Ages*. New York: Black Dog and Leventhal Publishers, 1990.

Grandin, Temple, and Catherine Johnson. *Animals in Translation: Using the Mysteries of Autism to Decode Animal Behavior*. New York: Scribner, 2005.

———. *Animals Make Us Human: Creating the Best Life for Animals*. Boston: Houghton Mifflin Harcourt, 2009.

Hardeman, Nicholas P. *Shucks, Shocks, and Hominy Blocks*. Baton Rouge: Louisiana State University Press, 1981.

Harris, Jessica B. *High on the Hog: A Culinary Journey From Africa to America*. New York: Bloomsbury USA, 2011.

Heurgon, Jacques. *Daily Life of the Etruscans*, translated by James Kirkup. London, UK: Phoenix Press, 2002.

Horowitz, Roger. *Putting Meat on the American Table: Taste, Technology, Transformation*. Baltimore, MD: The Johns Hopkins University Press, 2006.

Hudson, John C. *Making the Corn Belt: A Geographical History of Middle-Western Agriculture*. Bloomington: Indiana University Press, 1994.

Hurt, R. Douglas. *American Agriculture: A Brief History* (revised edition). West Lafayette, IN: Purdue University Press, 2002.

Jones, Evan. *American Food: The Gastronomic Story* (third edition). Woodstock, NY: The Overlook Press, 2007.

Jones, Martin. *Feast: Why Humans Share Food*. New York: Oxford University Press, 2007.

Jordan, William Chester. *Europe in the High Middle Ages*. New York: Viking/Penguin Group, 2001.

Kaminsky, Peter. *Pig Perfect: Encounters with Remarkable Swine and Some Great Ways to Cook Them*. New York: Hyperion, 2005.

Kiple, Kenneth F., and Kriemhild Coneé Ornelas, editors. *The Cambridge World History of Food*. Cambridge, UK: Cambridge University Press, 2000.

Kraig, Bruce. *Hot Dog: A Global History*. London, UK: Reaktion Books Ltd., 2009.

Kurlansky, Mark. *Salt: A World History*. London, UK: Random House, 2002.

Lauck, Jon K. *The Lost Region: Toward a Revival of Midwestern History*. Iowa City: University of Iowa Press, 2013.

Laudan, Rachel. *Cuisine and Empire: Cooking in World History*. Los Angeles: University of California Press, 2013.

Lewis, Celia. *The Illustrated Guide to Pigs: How to Choose Them, How to Keep Them*. New York: Skyhorse Publishing, 2011.

Long, Lucy M. "Food," in *The Greenwood Encyclopedia of American Regional Cultures: The Midwest*. Westport, CT: Greenwood Press, 2004.

Mallory, James, and Douglas Q. Adams, editors. *Encyclopedia of Indo-European Culture*. Chicago: Fitzroy Dearborn Publishers, 1997.

McGee, Harold. *On Food and Cooking*. New York: Collier Books Macmillan Publishing Company, 1984.

Mertz, Barbara. *Red Land, Black Land: Daily Life in Ancient Egypt* (revised edition). New York: William Morrow division of Harper Collins, 2008.
Miller, Adrian. *Soul Food: The Surprising History of an American Cuisine, One Plate at a Time*. Chapel Hill: University of North Carolina Press, 2013.
Mizelle, Brett. *Pig*. London, UK: Reaktion Books Ltd., 2011.
Morley, Neville. "The Profits of Empire," in *Cambridge Illustrated History: Roman World*, edited by Greg Woolf. Cambridge, UK: Cambridge University Press, 2003.
Nemet-Nejat, Karen Rhea. *Daily Life in Ancient Mesopotamia*. Westport, CT: Greenwood Press, 1998.
Ogle, Maureen. *In Meat We Trust: An Unexpected History of Carnivore America*. New York: Houghton Mifflin Harcourt, 2013.
O'Neill, Molly, editor. *American Food Writing: An Anthology with Classic Recipes*. New York: Literary Classics of the United States, 2007.
Pacyga, Dominic A. *Slaughterhouse: Chicago's Union Stock Yard and the World It Made*. Chicago: University of Chicago Press, 2015.
Puckett, Susan. *A Cook's Tour of Iowa*. Iowa City: University of Iowa Press, 1988.
Rogers, Katharine M. *Pork: A Global History*. London, UK: Reaktion Books, 2012.
Root, Waverley. *Food: An Authoritative and Visual History and Dictionary of the Foods of the World*. New York: Smithmark Publishers, 1996.
Rösener, Werner. *Peasants in the Middle Ages*, translated and with foreword by Alexander Stützer. Urbana: University of Illinois Press, 1992.
Shortridge, James R. *The Middle West: Its Meaning in American Culture*. Lawrence: University Press of Kansas, 1989.
Slade, Joseph W., and Judith Yaross Lee, editors. *The Greenwood Encyclopedia of American Regional Cultures: The Midwest*. Westport, CT: Greenwood Press, 2004.
Smith, Andrew F. *Eating History: Thirty Turning Points in the Making of American Cuisine*. New York: Columbia University Press, 2009.
———, editor. *The Oxford Companion to American Food and Drink*. New York: Oxford University Press, 2007.
———. *Starving the South: How the North Won the Civil War*. New York: St. Martin's Press, 2011.
Sponenberg, D. Phillip, Jeannette Beranger, and Alison Martin. *An Introduction to Heritage Breeds: Saving and Raising Rare-Breed Livestock and Poultry*. North Adams, MA: Storey Publishing, 2014.
Standage, Tom. *An Edible History of Humanity*. New York: Walker Publishing Company, Inc., 2009.
Stevens, Jolene. *Pigs! Lifting Mortgages, People, and Communities: A History of the Pork Industry in Lyon, Plymouth, and Sioux Counties in Iowa*. Sioux Center, IA: Dordt College Press, 2013.
Tannahill, Reay. *Food in History*. New York: Crown Trade Paperbacks, 1988.
Teuteberg, Hans J. "The Birth of the Modern Consumer Age: Food Inventions from 1800," in *Food: The History of Taste*. Berkeley: University of California Press, 2007.

Vehling, Joseph Dommers, editor and translator. *Apicius: Cookery and Dining in Imperial Rome*. New York: Dover Publications, Inc., 1977. (Republication of the 1936 edition published by Walter M. Hill in Chicago.)
Villas, James. *American Taste: A Celebration of Gastronomy Coast to Coast*. Guilford, CT: Lyons Press, 2010.
Watson, Lyall. *The Whole Hog: Exploring the Extraordinary Potential of Pigs*. Washington, DC: Smithsonian Books, 2004.
Weinzweig, Ari. *Zingerman's Guide to Better Bacon: Stories of Pork Bellies, Hush Puppies, Rock 'n' Roll Music, and Bacon Fat Mayonnaise*. Ann Arbor, MI: Zingerman's Press, 2009.
Willard, Pat. *America Eats!: On the Road with the WPA—the Fish Fries, Box Supper Socials, and Chittlin' Feasts that Define Real American Food*. New York: Bloomsbury, 2008.
Wiseman, Julian. *The Pig: A British History*. London, UK: Gerald Duckworth and Co. Ltd., 2000.
Wolmar, Christian. *The Great Railroad Revolution: The History of Trains in America*. New York: PublicAffairs, 2012.
Wood, Jacqui. *Prehistoric Cooking*. Pleasant, SC: Tempus Publishing, 2003.
———. *Tasting the Past: Recipes from the Stone Age to the Present*. Stroud, Gloucestershire, UK: The History Press, 2009.

Historic Cookbooks

Baptist Ladies' Aid Society. *The Baptist Ladies' Cook Book*. Monmouth, IL: Republican Print Company, January 1, 1895.
The Calumet Cook Book. Chicago: The Calumet Baking Powder Company, 1916.
Collins, Mrs. A. M. *The Great Western Cook Book or Table Receipts, Adapted to Western Housewifery*. New York: A. S. Barnes & Company, 1857.
The Ladies of the First Presbyterian Church of Dayton. *Presbyterian Cook Book*. Dayton, OH: Oliver Crook & Co., Printers, 1873.
Ladies of Plymouth Church, Des Moines, Iowa. "76." *A Cook Book*. Des Moines, IA: Mills & Company, Printers and Publishers, 1876.
Ladies of the Westminster Presbyterian Church, Keokuk, Iowa. *Cookbook of the Northwest*. Chicago: J. L. Regan Printing Co., 1887.
Owen, Mrs. T. J. V. *Mrs. Owen's Illinois Cook Book*. Springfield, IL: John H. Johnson, Printer, 1871.
Wilcox, Estelle Woods. *Buckeye Cookery and Practical Housekeeping*. Minneapolis, MN: Buckeye Publishing Company, 1877.

"Human Resources"

Birkenholz, Ron, communications director, Iowa Pork Producers Association.
DeKryger, Malcolm, president, Belstra Milling Company.

Dorsch, Megan, marketing manager, Nueske's Applewood Smoked Meats.

Drier, Carolyn, third-generation owner, National Historic Site Drier's Meat Market.

Eckhouse, Herb, co-founder and co-owner, La Quercia Handcrafted Cured Meats.

Goldwyn, Craig "Meathead," author of *Meathead: The Science of Great Barbecue and Grilling* and creator of AmazingRibs.com barbecue website.

Greenway, Brad and Peggy, pork producers, Greenway Pork Farms.

Hammond, David, dining and drinking editor, *Newcity Magazine*; columnist, *Wednesday Journal*, oakpark.com; commentator, Rivet News Radio; contributor, *Chicago Tribune* and *Chicago Sun-Times*; lead moderator, LTHForum.com.

Henry, Steven, veterinary clinician specializing in swine and population medicine, author, and international swine medicine consultant.

Hildebrandt, Dirk, historic farmer in charge of the agricultural program at Old World Wisconsin, a six-hundred-acre outdoor museum of Wisconsin immigrant architecture and living.

Hofer, Shorty, owner and operator, Shorty's Locker.

Kahan, Paul, award-winning chef, restaurateur, partner in One Off Hospitality Group.

Kelly, Megan, swine technician.

Nueske, Tanya, chief executive officer, Nueske's Applewood Smoked Meats.

Polcyn, Brian, award-winning chef, educator, and author (*Charcuterie: The Craft of Salting, Smoking, and Curing* and *Salumi: The Craft of Italian Dry Curing*).

Price, Max, archaeologist and lecturer on archaeology at Massachusetts Institute of Technology.

Robinson, Nate and Lou Ann, pork producers, owners Jake's Country Meats. http://www.jakescountrymeats.com/.

Ropp, Ken, farmer and cheese maker, Ropp Jersey Cheese. http://www.roppcheese.com/.

Schieck, Sarah, Swine Extension Educator, University of Minnesota Extension.

Stoddard, Elisabeth, Switzer Environmental Fellow, assistant teaching professor in Undergraduate Studies and Environmental Studies, co-director of the Center for Sustainable Food Systems at Worcester Polytechnic Institute in Worcester, MA.

Underly, Kari, author, master butcher, educator, and principal, Range, Inc., a Chicago-based meat marketing, training, and education firm. http://www.rangepartners.com/.

Walker, Paul, environmental management consultant, professor emeritus, Illinois State University, recently retired investigator, Livestock and Urban Waste Research Team, and program coordinator Swine Waste Economical and Environmental Treatment Alternatives.

Weiss, Judi Frye, farmer, owner Back Home Farms.

Wells, Carolyn, co-founder and executive director, Kansas City Barbecue Society.

White, Alice Mae, pig farmer's daughter.

Zatkoff, Matthew, filmmaker and charcuterie hobbyist.

Zeder, Melinda A., senior research scientist, Program in Human Ecology and Archaeobiology; curator, Old World Archaeology, Department of Anthropology, National Museum of Natural History, Smithsonian Institution.

Zurer, Seth, co-founder, Baconfest Chicago. http://baconfestchicago.com/.

On Location

Baconfest, Chicago, IL
Belstra Milling Company, DeMotte, IN
Chicago History Museum, Chicago, IL
Corn Palace, Mitchell, SD
Drier's Meat Market, Three Oaks, MI
Fort Wayne History Center, Fort Wayne, IN
Greenway Pork Farm, Mitchell, SD
The Henry Ford Museum/Greenfield Village, Dearborn, MI
Hormel Historic Home, Austin, MN
Iroquois Valley Swine Breeders, Fair Oaks, IN
Jake's Country Meats, Cassopolis, MI
Living History Farms, Urbandale, IA
Mower County Historical Society, Austin, MN
National Road/Zane Grey Museum, Norwich, OH
Nueske's Applewood Smoked Meats, Wittenberg, WI
Old World Wisconsin, Eagle, WI
One Off Hospitality, Chicago IL
Oriental Institute, University of Chicago, Chicago, IL
Pig Adventure at Fair Oaks Farms, Fair Oaks, IN
Publican Quality Meats, Chicago, IL
Red, White, and Bar-B-Q, KCBS-sanctioned Championship Contest, Westmont, IL
Ropp Cheese, Normal, IL.
Schoolcraft College, Livonia, MI
Shorty's Locker: Custom Butchering & Grocery, Mitchell, SD
Wisconsin State Fair, West Allis, WI
Plus travels in Asia, Europe, the Middle East, and North and South America, in places that love pigs and those that forbid them (or merely hold them in contempt).

Online Resources

Encyclopedia of Chicago History: http://www.encyclopedia.chicagohistory.org/.
Encyclopedia of the Great Plains, University of Nebraska, Lincoln: http://plainshumanities.unl.edu/encyclopedia/.
Food and Agriculture Organization of the United Nations: http://www.fao.org.
The Livestock Conservancy: http://www.livestockconservancy.org/.
U.S. Department of Agriculture Census of Agriculture: http://www.agcensus.usda.gov/.
U.S. Department of Agriculture Economic Research Service: http://www.ers.usda.gov/.

Index

Page numbers in *italics* refer to photos.